2019 ENTERPRISE TECHNOLOGY BUYER'S GUIDE

2019
ENTERPRISE TECHNOLOGY
BUYER'S GUIDE

Ian Campbell · Nucleus Research

Published by Nucleus Research
ISBN 978-1793930040
First printing: January, 2019

TABLE OF CONTENTS

TABLE OF FIGURES

FOREWORD

This Buying Guide represents the collective work of Nucleus Research analysts across all markets from 2018. It is designed as a guide to those evaluating software for purchase to help show the financial value that such software can bring to their companies. All research is performed and published independently of vendors and is not an explicit endorsement of any one vendor, nor does it claim that any one product will provide the highest value in a given business situation. In addition, full versions of the included matrices and case studies are available for further review from our research library.

With over 700 ROI case studies to date, Nucleus continues to prioritize customer feedback and real customer use cases above all else. With every year of case studies, value matrices, and other research, we engage with more customers and more vendors to better understand what is important for businesses to grow. We hope that this book can serve as a guide to companies making their vendor short-lists, narrowing down the winner of a close proposal process, or re-justifying their ongoing vendor subscriptions.

CHAPTER 1

INTRODUCTION

If you bet your business on blockchain, invested heavily in Internet of Things (IoT) or doubled down on wearables in 2018, it's time to reconsider how you make tech decisions. To be honest, it's hard not to get excited by the trends that grab the most coverage and generate all the buzz. Sadly, history shows us that major tech trends often don't pan out as predicted. Augmented reality, NFC chips and Google Glass were all on the verge of radically transforming the technology landscape—until they didn't.

The problem is that news coverage is often based on what is interesting and flashy. What grabs eyeballs and even more importantly, sells ad dollars. Increasingly, visuals and graphic context drive what gets covered. As a result, our collective tech media presents us with fun articles on conceptual self-driving vehicles accompanied by slick videos, all-things robotic and every single miniscule update to the iPhone.

Meanwhile, there are fewer and fewer articles on how Enterprise Resource Planning (ERP) can power up productivity. We also don't see very much coverage of Human Capital Management (HCM), Workforce Management (WFM) and Talent Management apps, despite the great need to find top talent in the current job market. Analytics—once the tech press darling—has been sidelined by Artificial Intelligence (AI) and its mostly highly futuristic applications. Even Customer Relationship Management (CRM), the steady performer of enterprise applications, has seen less press attention as everyone gravitates to the quixotic shiny objects. Gadgets, Instagram layouts and anything that "drives itself."

Yet organizations are tasked with making sound technology decisions that optimize business for 2019 and beyond. Following the latest consumer crazes is wasteful and even dangerous. Last year I noted that tech decisions are more critical to the success of your business than ever before in history. That statement is even more true today. Technology is becoming increasingly more central to nearly every business and is strategically critical for the enterprise.

It's one of the biggest tech trends of our lifetime, where the importance of technology and value it can deliver is growing at a seemingly exponential rate. No longer just the underlying infrastructure that enables networking in the workplace, technology is now at the very core — the business central nervous system, if you will. Making the right decisions today requires a deeper look at the technologies that can advance a company, well beyond media hype and social buzz.

It's All About Value!

Good business decisions are all about value. For technology, we determine that based on three factors. **Functionality** determines the capabilities an application or solution can deliver. Does that CRM app enable micro-coaching? Can this corporate performance management (CPM) app pull needed data from the ERP system? It's all about what a solution can do.

Usability identifies ease-of-use and how many people within the organization can actually take advantage of the application. Traditionally, complex applications such as Supply Chain were only accessed by a small number of employees. Broad usability was less important. But as we integrate applications, the need for more people across multiple teams becomes a priority. Usability is increasingly important for solutions to deliver value.

Cost is the third consideration for value. Can one application deliver comparable functionality and usability for a lower price? Does one solution offer other cost advantages over another, such as lower maintenance? While cost is very important, it's better to assess applications for the functionality and usability a business requires first, then do a cost comparison on the short list to zero in on the best choice.

Proven ROI and Success

While functionality, usability and cost can determine the value an application can deliver, ROI case studies illustrate how other organizations have already derived value. Demonstrating a blueprint for success with a numbers-based validation plus informative lessons-learned and best practices are contained within ROI case studies to provide more insight.

Nucleus Research has developed a visual grid that helps companies quickly understand how vendors rank in both functionality and usability. Called the Value Matrix, we help numerous business make better decisions. This book includes our Value Matrix for each business solution, along with illustrative ROI case studies to help convey which applications deliver the most value.

With this book and our tools, we help you cut through the hype and ignore the overstated promises of "cool" shiny objects to instead focus on tangible solutions that can help your company now. With tools based on real-world data and successful customer cases that make your own success more likely.

Customer Relationship Management Value Matrix

Dot indicates current position. Arrow indicates trend in 2019 relative to others in the market.

CHAPTER 2
CUSTOMER RELATIONSHIP MANAGEMENT

Vendors on the Matrix are compared based on the relative usability and functionality of their respective products compared to competitors and the overall market. As the core functional capabilities of marketing, sales, and service automation have matured, vendors are investing in four main areas:

Artificial intelligence (AI) and analytics. Embedded analytics and some AI capabilities have become table stakes for CRM competition today—even if users may be slower to adopt in some areas (such as sales force automation) than vendors would like. We expect continued investment here as vendors refine their delivery of AI to enable model-driven data cleansing and more transparent (and, thus, credible) models for prediction.

Integration. With most organizations having grown separate sales, marketing, and service technology functions (and, in many cases, multiple applications or instances in each core CRM area), disparate data is increasingly a challenge for providing any real-time understanding of a complete customer history. Vendors are making investments in integration both organically and by acquisition to help customers retain their existing investments while taking advantage of cross-core opportunities.

Usability. Most Leaders in the Matrix are making usability enhancements to their products on an ongoing basis, either through complete UI overhauls, subtle UI enhancements, or additional embedded intelligence and coaching features that enable users to automate more of their daily tasks and follow best practices.

Edge capabilities. Seeking to extend their overall footprint and make their solutions stickier, vendors are continuing to invest in edge CRM capabilities—such as field service automation, configure price quote (CPQ), and enhanced analytics—that drive greater ROI for customers.

· · · · ·

Acumatica

Acumatica is primarily an ERP provider; however, it is included in this Matrix because Nucleus has found that customers can rely solely on Acumatica technology to meet their CRM needs. Acumatica CRM and ERP solutions are both integrated on the same backend database which ensures accurate records across departments and no double-entry of data. Acumatica solutions are most commonly used by small and medium-sized business (SMB) customers primarily in the retail, wholesale distribution, service, manufacturing, construction, and technology industries. Acumatica offers solutions for sales automation, marketing automation, service and support automation, and customer self-service as well as e-commerce through its Magento partnership.

Acumatica demonstrates its commitment to delivering a quality CRM solution with regular product improvements and is poised to move right in future Matrices as it shows further efforts toward delivering additional advanced functionality to users.

Bpm'online

Bpm'online offers a business process management (BPM) platform that can be leveraged to manage processes throughout the business, including CRM. The product is improved at a rapid pace with monthly feature updates. Recently announced strategic partnerships demonstrate that the company is experiencing growth, looking to expand its customer base, and add new integrations and capabilities to its platform.

Bpm'online remains firmly positioned in the Leaders quadrant because of the high usability of the solution and its significant investment in improving functional capabilities such as predictive analytics, AI, and data management.

ROI Case Study: Bpm'online

Hershey's Ice Cream • ROI: 135%

Hershey's Ice Cream deployed Bpm'online to modernize its sales and account management practices and provide a consistent source of data for analysis and decision making. Moving from a largely paper-based system to Bpm'online and providing sales managers in the field with tablet-based access enabled the company to increase productivity, identify opportunities for sales performance improvements more quickly, and redeploy three sales support resources by automating data entry and analysis.

CUMULATIVE NET BENEFIT

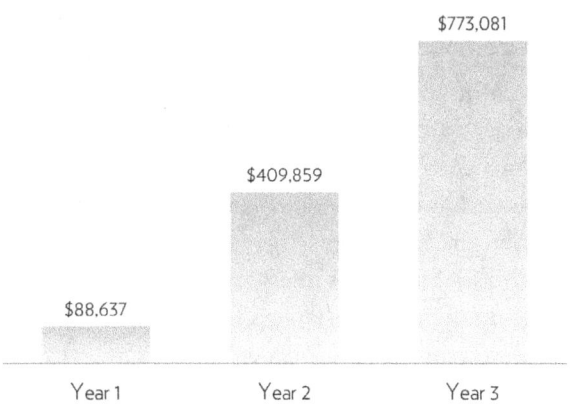

		$773,081
	$409,859	
$88,637		
Year 1	Year 2	Year 3

THE COMPANY
The Hershey Creamery Company was founded in 1894 by Jacob Hershey and his four brothers on a farm in Lancaster Pennsylvania. In the past 124 years, the family-owned company has grown from a small farmhouse operation to more than 40 distribution sites with sales of over $140M a year. With their involvement beginning in 1924, the Holder family has owned and operated Hershey's Ice Cream since the early 1960s. The company, now based in Harrisburg, Pennsylvania, is on the top 100 list of US

ice cream manufacturers, selling more than 120 flavors of hand-dipped ice cream, sherbet, and frozen yogurt, as well as novelty and packaged desserts. The company sells through dealers as well as directly to schools and institutions.

KEY BENEFIT AREAS

Deploying Bpm'online to support its sales operations enabled Hershey's to streamline data entry, increase sales productivity and effectiveness, and drive more data-driven decision making. Key benefits of the project included:

- **Increased sales productivity.** Because sales can access information on all their accounts via a tablet, they can plan their account visits with complete visibility into account health, location, and order history. They can also easily update their account activity from the road, saving time while increasing the timeliness and detail of account information.
- **Reduced administrative overhead.** Because sales reps can now enter information directly and reports and dashboards are automatically updated with real-time data, Hershey's has been able to redeploy three administrative staff members to other work activities while significantly reducing paper and printing costs.
- **Reduced travel costs.** Greater visibility into account activity and location for both field sales staff and managers enables them to plan their account visits to optimize visit time and minimize travel time, increasing "face time" with accounts while reducing overall fuel and travel costs.
- **Improved sales management.** Before Bpm'online, managers had to wait — sometimes months — for sales productivity data to be available. Managers now have complete visibility into sales activities as they happen, enabling them to be more proactive and data -driven in providing sales coaching and helping field sales to optimize their account management activities. They can

also segment productivity by daily and weekly numbers as well as by the type of customer visited (such as "assisted living facility" or "convenience store"). Over time, the company expects that this will enable managers to onboard new field staff more rapidly by providing better guidance on account prioritization and sales offers.

BEST PRACTICES
Becoming a more data-driven business has enabled Hershey's to become much more proactive in their coaching to ensure that field sales reps are effective in both how they manage their time and how they share data about accounts. A key part of this data-driven approach was an easy-to-access and use tablet application that makes it easier for them to take on new accounts and plan their activities while encouraging greater data capture.

· · · · ·

Hubspot

HubSpot is a provider of cloud-based sales and marketing automation solutions. It has historically catered to SMBs, but recent activity suggests it is looking up-market to grow its customer base and mature its products. For paying customers using its marketing automation solution, HubSpot offers its sales force automation (SFA) tools for free. It recently announced the launch of enterprise-level products that were released on September fifth.

With its focus turned up-market on enterprise customers, HubSpot has made some improvements to the functionality of its product with emphasis placed on enabling more advanced analytics in its enterprise-class products. However, it lacks the resources of many leaders to invest in keeping pace on functionality.

Indegene

Indegene offers customer engagement for life sciences with coverage across all sectors of the healthcare industry. Through a strategic partnership with Microsoft, Indegene Omnipresence, leverages Microsoft technology such as

Dynamics 365 CRM and Azure data and analytics to deliver advanced functionality specific to the healthcare vertical. Indegene differentiates itself by enabling advanced functionalities with a particular focus on industry-specific advanced analytics, cognitive capabilities, and natural language processing (NLP). Other features include integration with social platforms such as LinkedIn (through Microsoft Dynamics) for customer engagement and creation of custom AI-powered bots to handle repeatable tasks or customer queries, including medical inquiries.

Indegene positions itself as a cross-functional organization that acts as a digital transformation business partner with healthcare-specific best practices and content creation to help its partner organizations grow. Additionally, its all-inclusive pricing strategy makes it an attractive option for organizations looking for a different business case.

As a young vendor still growing its customer base and expanding the platform capabilities, Indegene is positioned as a Core Provider in this Matrix. If it continues the pace of expected development in additional transactional capabilities, Indegene is on track to move into the Facilitator quadrant in the future.

Infor

The Infor Customer Experience (CX) Suite includes Infor Configure Price Quote, Infor Contract Lifecycle Management, Infor CRM, Infor Interaction Advisor, Infor Marketing Resource Management, Infor Omni-channel Campaign Management, Infor Rhythm, Infor Sales Intelligence for CRM, and marketing automation through its partnership with Marketo.

Infor's position is slightly degraded in the Leaders quadrant since the last edition of the Value Matrix because the pace of enhancements to product usability and functionality is lagging behind other players in the market. However, we expect announcements of a significant reengineering effort in the product in the near future that will likely drive a positive change in positioning in the next Matrix.

Infusionsoft

The Infusionsoft platform brings CRM together with marketing automation, e-commerce, and payments for 180,000 small business users. By organizing all customer interactions in one place and making it easy to connect

Infusionsoft with thousands of other apps, small businesses can automate key customer activities and work intelligently to deliver more personalized service and close more business. This year the company announced a simplified version of its platform designed to meet the needs of solopreneurs and the broader small business market.

IQVIA

IQVIA differentiates itself from other CRM vendors with its Orchestrated Customer Engagement (OCE) platform, described by IQVIA as a new category of technology that allows businesses to leverage an "orchestrated commercial model" across all of its customer-facing functions. This is a "platform-of-platforms" strategy that combines the company's expertise in the health and life sciences with industry-standard platforms, partnerships with best-of-breed component and technology vendors, and disruptive technologies like AI.

IQVIA's position in the Facilitator quadrant is degraded slightly from the last publication of the Matrix since the lack of additional product improvements show that it is lagging behind competitors. However, we expect additional organic investment (beyond its partner strategy) will pay off in future Matrix positioning as customers are fully deployed.

Maximizer Software

Maximizer CRM delivers industry-specific sales, marketing, and customer success capabilities to SMBs including dedicated platforms for financial services and manufacturing. With industry-specific versions available, the need for custom configuration is decreased which allows Maximizer to offer complete CRM solutions at a lower cost. Since the last Matrix, there haven't been many significant product releases or feature upgrades; however, the new 2019 version delivers total pipeline management with a new lead module and a dedicated native mobile Sales app available for iOS and Android.

This year, Maximizer announced a new integration with QuoteWerks, a leading sales quoting and proposal software provider. The partnership will allow customers of both companies to accelerate the process of creating quotes, proposals, and sales opportunities by automatically linking data between the two applications.

This partnership shows that Maximizer is looking to expand the breadth of its business tools to reach new customers and deliver an improved customer experience to its existing users. Still, Maximizer has lagged behind the market in differentiating areas like AI and low-code development options.

Microsoft

With Microsoft Dynamics 365, Dynamics CRM and enterprise resource planning (ERP) capabilities are unified on one platform to provide an integrated suite of solutions for sales, marketing, service, finance, operations, and talent management that is also seamlessly connected with Office 365. Like its competitors, Microsoft is focused on augmenting its enterprise tools with AI using Azure, and by using PowerBI and PowerApps, customers have access to no-code, low-code, and full development capabilities for custom analyses.

Microsoft announced new Dynamics 365 AI offerings for Sales, Customer Service, and Marketing to deliver out-of-box insights by unifying data and applying advanced intelligence to it. Microsoft also unveiled the first mixed reality applications for business, leveraging HoloLens technology to offer Dynamics 365 Remote Assist and Dynamics 365 Layout. Dynamics 365 Remote Assist is an application allowing experts to connect with employees for remote collaboration with heads-up, hands-free video calling, image sharing, and mixed reality annotations. Dynamics 365 Layout is a subscription application that imports 3D models to create and edit room and area layouts at real-world scale.

With improvements to the UI, automation, and custom configuration using code and code-free environments, Microsoft has kept pace with market leaders in usability. It is outstripping the market in functionality with increasing investment in AI and analytics, improved agent and supervisor experience, and edge technologies like mixed reality.

 ## ROI Case Study: Microsoft Dynamics CRM

MacDonald-Miller Facility Solutions • ROI: 282%

MacDonald-Miller Facility Solutions (MMFS) deployed an integrated solution of Microsoft Dynamics 365 with Azure IoT Hub to improve operational efficiency, centralize company information and processes, and to modernize as an early adopter of Internet of Things (IoT) technology. MMFS was able to modernize its field service operation to one of proactive maintenance, reduce the time-to-completion for service calls by nearly two weeks, increase profits, and differentiate itself as an industry leader in smart building construction, optimization, and maintenance.

CUMULATIVE NET BENEFIT

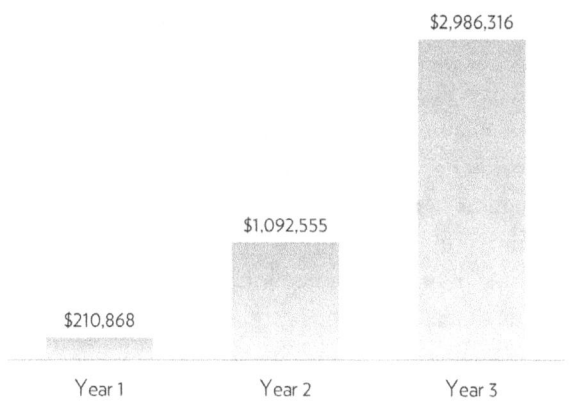

	Year 1	Year 2	Year 3
	$210,868	$1,092,555	$2,986,316

THE COMPANY

MMFS is a leading full-service, design-build contractor with three primary business arms specializing in new construction, facility service and maintenance, and building performance optimization. Headquartered in Seattle, WA with 10 locations across Washington and Oregon, MMFS is an industry leader in the design, construction, and modernization of "smart" buildings to minimize building energy usage with the goal of cost savings for owners and a "greener" planet for all.

KEY BENEFIT AREAS

Deploying Dynamics 365 and Azure IoT Hub allowed MMFS to modernize and grow its business by facilitating collaboration across business units and departments and optimizing field service operations by leveraging IoT. Key benefits of the project included:

- **Increased field service efficiency.** Because of field service automation and the integration of IoT data, MMFS is able to manage more buildings and facilities without a correlated increase in field staff.
- **Increased profits.** The project enabled MMFS to increase profits through improved marketing, increased lead capture by technicians in the field, and cross-departmental transfer of leads. The company has also gained new opportunities through its differentiated IoT service capabilities.
- **Increased visibility.** MMFS was able to use analytics tools such as search engine optimization (SEO) and keyword analysis to track and improve marketing campaigns in real time. The adoption of Dynamics by the marketing team has recouped over 58 percent of the total deployment costs in additional profits from Web-generated leads over the past 6 months.

BEST PRACTICES

Integrating analytics with real-time sensor data from smart buildings differentiates MMFS as the industry leader in real-time, data-supported building optimization. As IoT is still a developing field, having Microsoft as a partner with a reputation for industrializing nascent technologies lent MacDonald-Miller both validity and expertise in the deployment and has allowed it to redefine its business model from one of reactive service to one of proactive engagement.

Realizing that adoption was critical to achieving value from the deployment, MMFS knew it needed an internal transition strategy that included both user training and management support. To incentivize the migration onto the new Dynamics platform, it set up an internal adoption leaderboard. Using the leaderboard to create individual and team usage benchmarks, managers were able to identify the different classes of users. Frequent and skilled users scored higher and were commended while the more reluctant adopters were supported, resulting in a faster and more well-accepted migration to Dynamics.

. . . . .

NexJ

NexJ is a customer process management-focused CRM solution for the financial services and insurance industries, providing client onboarding, Know Your Customer (KYC), and customer data and analytics solutions. Although NexJ lacks the access to capital or deep developer bench of larger, more entrenched CRM vendors, its track record of investment in AI and other edge technologies justify its position as an Expert in this Matrix.

 ## ROI Case Study: NexJ

Financial Services Company • ROI: 42%

A global financial services firm deployed NexJ to replace its aging Oracle Siebel applications which were no longer effective tools for the sales team. Nucleus found that the project enabled the bank to improve technology management by moving to a modern platform that could support new communication channels with embedded analytics and insights. With the new application, the bank was able to increase data capture by 25 percent. Key benefits included reduced technology costs, increased sales and manager productivity, reduced administrative overhead, and increased customer satisfaction.

THE COMPANY
The bank profiled in this case study is one of the largest banking and financial services organizations in the world, operating more than 7,000 offices in Europe, the Americas, the Middle East, Africa, and the Asia-Pacific region. The

division profiled in this case study employs more than 1,500 sales people.

CUMULATIVE NET BENEFIT

$9,533,304

($4,360,386)

($12,612,231)

Year 1   Year 2   Year 3

KEY BENEFIT AREAS

Moving to NexJ has enabled this financial institution to provide sales people and managers with capabilities for data-driven decision making, reduce the burden of manual data entry, and support more intelligent interactions with clients with next-best action recommendations gleaned from both internal and external data. Key benefits of the project include:

- **Improved technology management.** Eliminating Siebel enabled the company to cut $3.8 million in annual license maintenance as well as the IT support costs of maintaining Siebel, freeing up IT time for more strategic activities.
- **Reduced administrative overhead.** The administrative staff team that was previously tasked with entering sales people's notes into Siebel have been redeployed, as now sales has a usable application they can access in the field to update their own client records.
- **Increased sales and manager productivity.** Better access to information, integration with e-mail, and embedded analytics and dashboards enable sales and

management to spend less time on data entry and analysis and more time on managing client relationships.

- **Reduced cost of sales.** Better insight into account health has enabled sales to better prioritize its customer engagement visits and events, enabling them to reduce spending on travel and other client-facing expenses that are no longer needed to engage clients.

BEST PRACTICES

With any new application, getting users to trust the data is critical to adoption. The bank initially underestimated the time and effort that would be needed to scrub and operationalize the legacy data, and found that data discrepancies distracted users, initially, from the overall improved functionality and user experience. Additionally, time and expense devoted to data scrubbing and migration efforts should be expected by any company considering a migration of this type.

Recognizing that user adoption was going to be a key challenge for the team, the project team invested in bringing in users for testing and input early in the process who could then be used to champion the benefits of the new application within their own groups. This approach, coupled with Web-based and other training and marketing, helped users to understand the benefits of the new application to them and smoothed adoption.

· · · · ·

Oracle

Oracle Customer Experience (CX) Cloud includes functionality for sales, marketing (including loyalty management), service, configure price quote (CPQ), sales performance management (SPM), social, business-to-business and business-to-consumer commerce, and customer data management. Oracle's strong focus on data management capabilities, stemming from its time as primarily a database provider, is a strong differentiator for its CRM offering compared to other solutions.

Oracle has continued investing in AI-implementation in the Oracle CX Cloud to bring deeper insights and better recommendations to business users and unlock productivity gains with task automation.

Oracle is looking to encourage cloud adoption among its customers with its "Soar to the Cloud" initiative. It is an automated enterprise cloud application product that consists of automated tools and best practices for cloud transitions. It is currently only available for ERP deployments, but Oracle is investing internally to offer Soar for the CX Cloud in the future.

Due to its comprehensive end-to-end capabilities and data management expertise, Oracle remains positioned as a Leader in this Matrix with the highest rated functionality. This doesn't look to change as Oracle has embraced the migration to the cloud and growing customer preference for full-suite offerings.

Pegasystems

Pegasystems provides industry-specific CRM capabilities for sales, marketing, and customer service. Its primary areas of investment are AI and robotic automation, which are key differentiators for its new CRM product, Pega Infinity.

This year, Pega announced the replacement for Pega CRM Suite, Pega Infinity. This suite contains the marketing, sales, and service solutions along with Pega Customer Decision Hub to offer a single unified platform for CRM. Support for custom robotics is automatically embedded on the platform. Pega Self-Optimized Campaigns leverage AI to automatically identify the optimal audience and deploy the campaign in self-monitored "waves" to reach each customer in the most ideal way. Natural language processing (NLP) is built into email to identify content and sentiment within messages, and AI-powered OCRs scan and contextualize attached documents. With Pega Infinity, Pega is an early adopter of blockchain technology through an integration with Etherium, offering the solution to streamline the document-heavy "know your customer" onboarding process.

With these functional improvements, Nucleus sees Pega targeting growth and looking to expand its positioning in the CRM space. However, as a leading adopter of edge technology like AI, blockchain, and NLP, overall usability may suffer as customers find themselves unsure how to best leverage these new capabilities.

Sage

Sage CRM is offered as integrated, modular CRM, available as part of Sage Business Management solutions around the globe. Providing CRM functionality out-of-the-box, with rich configuration capability, the proposition is at a cost-effective price point, particularly for Sage ERP customers in the midmarket that require the core CRM functionality covering marketing, sales, and service functions within a typical business. Customers looking for next-generation features like AI and low-code frameworks can take advantage of the Sage integration with Salesforce via the app cloud.

Salesforce

Salesforce's CRM clouds include Sales Cloud, Service Cloud, Marketing Cloud, Community Cloud, Einstein Analytics, Industry Clouds (for financial services, healthcare, government, education, and non-profit), Philanthropy Cloud, Commerce Cloud, and Integration. Salesforce continues to encourage customers to adopt or transition to Lightning, its modern UI and development platform. With Lightning, customers have access to the most up-to-date features and innovations from Salesforce, such as Einstein, its AI. With Trailhead, an online learning and developer experience community, users can educate themselves and grow their skillset with Salesforce technology.

This September, Salesforce announced new CRM functionality to Sales Cloud and Pardot: High Velocity Sales for Sales Cloud includes sales cadences and work queues, Salesforce Billing enables customers to manage their Salesforce subscription directly within the Salesforce ecosystem. It now offers usage-based pricing, evergreen subscriptions that automatically renew, and flexible invoicing. Additional Einstein capabilities were implemented on Pardot such as Einstein Campaign Insights and Einstein Behavior Scoring.

Salesforce moves even further toward the top of the Leader quadrant of this edition of the Value Matrix due to its continued investment in improving product usability and functionality ahead of the pace of the market and similar competitors.

ROI Case Study: Salesforce

ICS+ • ROI: 942%

ICS+ deployed Salesforce Sales Cloud, Service Cloud, Community Cloud, Quip, and Inbox, to provide one integrated platform for all client-focused data and collaboration. Moving from NetSuite, e-mail, and a variety of spreadsheets and other tools enabled the company to streamline project management, increase data capture and collaboration, increase employee productivity, and accelerate collections by one-third while increasing client satisfaction.

CUMULATIVE NET BENEFIT

Year 1	Year 2	Year 3
$121,683	$262,558	$417,500

THE COMPANY

ICS+ specializes in building automation control systems, specializing in audio and video solutions for commercial properties such as hospitals, hotels, airports, and educational institutions, and for some large residential estates. Headquartered in Austin, TX, the company has been in business more than 10 years, growing as demand for custom video and audio installations in commercial properties has increased. Although the company has fewer than 10 employees, it completes roughly 100 to 115 client engagements each year.

KEY BENEFIT AREAS

Deploying Salesforce has enabled ICS+ to continue to grow its business while improving its margins by having greater visibility into day-to-day operations. Key benefits of the project included the following:

- **Increased productivity.** Integration of quip with Salesforce and providing clients with access to Quip has reduced the amount of e-mail traffic by an average of 20 percent, with more precise details being captured within one system of record.
- **Improved technology management.** Moving from NetSuite enabled the company to significantly reduce its annual software license subscription fees while reducing the overall time needed to support technology.
- **Change in working capital.** Because of more timely and accurate project management data within Salesforce, ICS+ can invoice and collect from customers faster, shortening its accounts receivables cycle by 15 days.
- **Increased sales.** Tracking opportunities within Salesforce has enabled the company to track and close jobs better so it can both be more selective about opportunities it wants to pursue and close more deals.

BEST PRACTICES

The low code and no code capabilities of Salesforce were an important factor in enabling ICS+ to make ongoing changes to its Salesforce footprint as its business needs evolved. Although the company has used some consulting services, it is able to perform many enhancements and custom modifications without the cost of outside help that was always needed when modifications to NetSuite needed to be made. It has also been able to leverage the Salesforce Success Community as a resource for support.

· · · · ·

SAP

This year, it was announced that SAP Hybris Cloud for Customer has been rebranded by SAP, with its component cloud products now included in the SAP C/4HANA suite. The C/4HANA suite consists of SAP Marketing Cloud, SAP Commerce Cloud, SAP Service Cloud, SAP Customer Data Cloud, SAP Sales Cloud (formerly SAP Hybris Revenue Cloud and SAP Hybris Cloud for Customer), and the recently acquired CallidusCloud portfolio.

The move serves to consolidate SAP's acquisition portfolio with existing database technologies in a new full-service CRM offering. SAP C/4HANA goes beyond traditional sales-focused CRM and connects the customer experience to the actual supply chain for improved communication, visibility, and transparency. The solution is natively integrated with SAP S/4HANA, its ERP suite, and supports machine learning capabilities from SAP Leonardo. SAP HANA Data Management Suite is included with C/4HANA to enable data-driven applications by providing one secure, unified location to orchestrate data.

We see SAP as a Facilitator on this Matrix, due in part to it embracing the market trend toward full suite products. Through complete integration with ERP software, SAP looks to deliver high cross-functional usability throughout the entire business, although its functionality remains lacking compared to some Leaders in terms of code-free object creation and advanced analytics.

Satuit Technologies

Satuit Technologies offers cloud-based and on-premise CRM solutions, built specially for financial services clients including private equity and hedge fund managers, wealth managers, institutional asset managers, and funds distribution. The solution is equipped with core CRM functionality and contains additional tools specific to the financial services industry to manage the extensive compliance and legal requirements characteristic to that industry.

Its most recent update includes improvements to data visualization, data handling, and analytics. Client and customer engagement metrics are enabled, and new pipeline visualization tools allow sales teams to strategically optimize their efforts to the most profitable accounts. Managers have full visibility into up-to-date activity records for each account. Functionally, Satuit included new tools to support data import and management, as well as additional analytics capabilities.

Satuit is positioned in the Leaders' quadrant of this matrix due to continued investment in its product to deliver improved usability and increased functionality to customers.

 # ROI Case Study: Satuit Technologies

CALCAP Advisors • ROI: 374%

CALCAP Advisors deployed Satuit to modernize and automate its investor management and communications processes and position itself for growth. We found that the Satuit project enabled the firm to increase the consistency and efficiency of communications, improve security, and provide clients with a professional self-service portal for improved transparency and better client reporting while increasing partner productivity by 2.5 percent.

CUMULATIVE NET BENEFIT

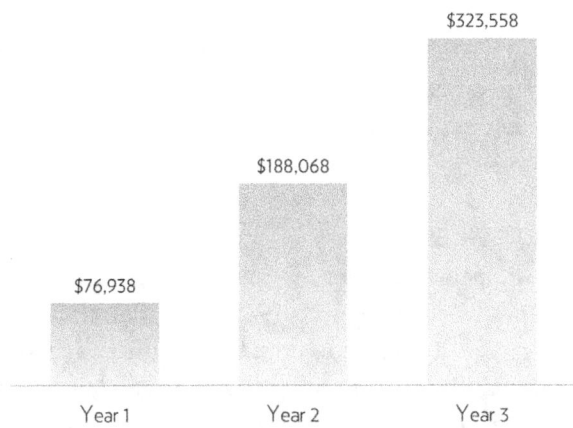

		$323,558
	$188,068	
$76,938		
Year 1	Year 2	Year 3

THE COMPANY
CALCAP is a boutique real estate investment and advisory firm based in Pasadena, California. Partnering with businesses, asset managers, real estate investors, and

developers, the company represents individual and institutional investors by strategically allocating capital across a multitude of real estate investments. Founded in 2008, the company has raised over $235 million in investor equity and is currently managing over $180 million in assets.

KEY BENEFIT AREAS

Moving to a single platform to consolidate account and investor information has enabled CALCAP to streamline and automate its financial reporting processes while providing around-the-clock secure access to clients via a secure portal. Key benefits of the project include:

- **Increased efficiency.** Previous manual processes of distributing information to investors have been eliminated, freeing up high-level staff's time for other client development and business efforts.
- **Increased visibility.** Having a single source of information for both partners and clients enables partners to focus on relationship building instead of more tactical questions from investors.
- **Increased competitiveness.** Having a secure, professional grade portal to show both investors and prospects enables CALCAP to compete more effectively against other investment firms that have invested significant resources in data delivery and self-service tools for investors.

BEST PRACTICES

CALCAP operates a unique business in the real estate sector, and the company recognized that to meet its growth targets it needed to move beyond the basic tools available via Microsoft Excel and Dropbox and present a more modernized digital presence to clients and prospects. Moving to Satuit's cloud application has positioned CALCAP to grow its investor base and thus the overall scale of its business without adding significant internal resources. Although the company will likely add resources for communications and client care as it grows, it can also take advantage of Satuit's data structures that are already customized for a parent-child relationships and other characteristics of CALCAP's business.

· · · · ·

SugarCRM

SugarCRM delivers a simple but highly-usable core CRM application with end-to-end functionality to support the entire business across sales, service, marketing, and customer support. Its Hint relationship intelligence tool delivers data-driven contextual insights to sales and service teams in the workflow. The product is designed with a focus on the user interface, with the belief that for business tools to be most effective, they should resemble the consumer applications that are so ubiquitous in everyday life to drive user adoption.

In August, it was announced that private equity firm Accel-KKR made a significant strategic investment in SugarCRM to enable its next phase of growth. Accel-KKR has a long track record of driving growth in portfolio companies. With the partnership, Sugar is looking to mature its capabilities in lead management and marketing automation, artificial intelligence and advanced analytics, and tools for customer support.

We place Sugar in the Facilitator quadrant of the Matrix due to its track record of prioritizing the usability of the platform instead of advanced functionality. This momentum is shifting as Sugar looks to continuously add new capabilities to the platform such as more sophisticated automation and improvements to the Hint relationship intelligence platform.

Veeva Systems

Veeva Systems is one of the few vendors in the Matrix that is specifically focused on one vertical: life sciences. Veeva Systems was founded on the Salesforce platform as a Salesforce partner and now delivers capabilities for clinical trial data management, content management, customer reference data, master data management, and other industry-specific capabilities critical to the life sciences industry. Veeva CRM products include Veeva CRM MyInsights for data visualization and Veeva CRM Engage Meeting and Webinar for remote interaction. Veeva updates its products every four months, in line with most other CRM vendors' release cycles.

Due to its construction on the Salesforce platform, significant investment in UI improvements, and additional vertical-specific capabilities for the health sciences field, Veeva moved into the Leaders quadrant of this Matrix.

Zoho

Zoho competes primarily in the SMB market with its low-cost all-in-one suite of applications for business, Zoho One. The suite offers cloud-based solutions for sales and marketing, e-mail and collaboration, business process management, finance, IT and help desk, and human resources in addition to integrated point applications for extended functionality. Since the last publication of the Matrix, Zoho has made aggressive moves to increase brand awareness and target up-market clients to grow its footprint.

By targeting up-market customers with an aggressive new marketing campaign and incorporating edge technologies like AI, Zoho is clearly developing its product to be more functional while being sure not to sacrifice the usability that attracted their core SMB customers in the first place. However, it is positioned as a Core Provider in this Matrix because the CRM-specific capabilities included in the Zoho One suite remain fairly basic.

Human Capital Management Value Matrix

Dot indicates current position. Arrow indicates trend in 2019 relative to others in the market.

CHAPTER 3

HUMAN CAPITAL MANAGEMENT

W ith leading HCM vendors delivering new capabilities several times a year, the competitive landscape continues to shift quickly. Part of Nucleus's four pillars of an organization, "people" controls one of the largest shares of budget. Payroll, talent retention, succession, and workforce management have a major impact on the bottom line today.

Leaders have set themselves apart by aggressively innovating in functionality around analytics and AI, real time prescriptive and predictive recommendations, and broad usability and accessibility. Major functional modules have now become table stakes, with leading vendors differentiating among themselves by providing integration collaboration and messaging.

Our analysis of trends among users shows that cloud-based technology will be the centerpiece of HCM, moving forward. Our analysis of users' experiences this past year shows that increasingly, organizations are turning to single suite applications as cross market integration becomes a greater business driver. Organizations need to be able to share and use data in real time across departments, which remains much easier to do in a single suite. Best-of-breed applications, particularly those vendors that do not have turn-key integration with larger solutions, will find it harder to stay in business.

More than 60 percent of customers are in the cloud. We found that those not in the cloud are either looking to move or are limited to on-premise solutions because of external factors such as unions. As a result, the majority of employers that still make do with on-premise solutions have very few options to stay there. Because of this, more than 80 percent of the on-premise customers that we interviewed are actively considering moving soon to the cloud. These realities explain related findings on planned spending in HCM. The data from this survey show all but a small handful of cloud-based users expect to increase spending per capita, per year, over the course of the next 12 months.

As human capital management (HCM) vendors and decision makers have accelerated their move to the cloud, vendors have focused on technological features, not value drivers. The debate around the cloud should be about the ability to deliver faster time to value and greater benefit over time with limited cost and disruption to end users. We have identified three key areas customers should consider: initial deployment and time to value, usability and accessibility, and the pace and impact of upgrades. Vendors that are able to show value in these three areas will dominate the market as on-premise deployments come to an end.

Embedded analytics and artificial intelligence (AI) are also becoming a must have for leading vendors. End users expect completely integrated analytics that is not only predictive, but prescriptive. The main analytics track in HCM is currently around employee flight risk and turnover prevention but is also expanding into areas like schedule optimization and succession planning. Analytics solutions can only be of value to end users if the data is continuously uploaded and available into a single offering.

· · · · ·

ADP

ADP offers three multitenant cloud solutions for HCM that cater to employers based upon size. ADP Run is intended for companies with 1 to 50 employees, Workforce Now for firms with 50 to 3,000 employees, and Vantage HCM is designed for organizations with over 3,000 employees. Each of these solutions offers a wide array of functionality. These include the ability to predict employee flight risk through ADP's Turnover Probability predictive model, enhanced EEO-1 compliance with Pay Equity Explorer's compensation analytics and benchmarking data capabilities, and collaboration with the U.S. Internal Revenue Service (IRS) on W-2 verification codes to reduce stolen income tax refunds.

ADP's acquisition of The Marcus Buckingham Company has enabled it to bolster its talent management capabilities with coaching and educational resources. ADP is an appropriate choice for global payroll due to its greater capacity to support this feature, relative to other vendors. Its wide degree of functionality, together with its usability shortcomings, justifies its placement within the Expert quadrant. ADP now offers Wisely Pay, a feature that enables clients to provide employees with multiple options to receive, spend, and manage money, financial management tools, and the ability to avoid paper check fees. ADP also partners with Microsoft Dynamics 365 Business Central to provide mutual clients with all in one business and HR functionality that includes finance, operations, sales, payroll, time and attendance, tax services, and benefits and talent management. The vendor also partners with collaboration hub Slack to enable mutual clients to provide their workers with access to HR and payroll information such as pay notifications and details, and time off balances and requests.

Ascentis

Ascentis is a cloud-based solution that spans core HR, payroll time and attendance, reporting, talent management, and talent acquisition. There is support for 360-degree assessments for reporting, and dashboards are available to conduct an analysis. The majority of this solution is accessible both conventionally and through its mobile UI. The solution enhances productivity by automating compliance tasks, such as the filing of ACA-pertinent information to the IRS.

Compliance functionality takes center stage within Ascentis's mobile ecosystem, with users speaking highly of it. The solution also differentiates itself from other applications with the ability of clients to purchase and deploy only the functionality they need. Clients can start working with any one of the Ascentis modules and migrate to other modules over time with no dependence on any specific module and no loss of functionality. Additionally, it is highly customizable with low maintenance integration between Ascentis solutions, other 3rd party HCM solutions and financial system platforms.

Ceridian

Ceridian built Dayforce from its origins in payroll and the rest of workforce management (WFM) into a full-breadth, full-fledged solution for HCM with functionality that now spans all of core HR, WFM, recruiting, onboarding, performance management, and compensation planning. Ceridian's large pool of HCM data means that the vendor is able to leverage its knowledge to aid in everything from initial deployment to predictive employee succession planning. Ceridian's single rules engine governs a single application which produces a single dataset. This architecture makes the definition of real-time processing of information possible. This architecture has set Dayforce apart from the competition and supports real time reporting.

Ceridian launched the Dayforce Software Partner Program (DSPP) last year to create a software partner ecosystem aimed at helping organizations easily connect other HCM-related solutions with Dayforce. The DSPP gives Ceridian software partners and application developers access to Ceridian's APIs, making data available across Dayforce and partner solutions. Contiguous and regular upgrades have now become a part of Ceridian's standard upgrade schedule. In addition to the twice-yearly major updates, Ceridian is providing smaller updates every six weeks. These updates go beyond fixing standard bugs and errors, to focus on adding incremental value to the end user's needs.

Ceridian is developing a voice-activated personal assistant that, using advanced analytics, will have the ability to recommend specific actions or transactions based on collective data and the experiences of others in similar situations. The assistant, which will work with smart speakers such as Amazon Alexa, Google Home, and Microsoft Cortana, will enable managers, workers, and administrators to complete tasks across all areas of HCM. Ceridian also offers an alliance that will seamlessly connect Ceridian's Dayforce with Microsoft Dynamics 365. Through this collaboration, Ceridian will provide the Dayforce payroll solution to Dynamics 365 customers in a single product offering. Backed, supported and productized by Microsoft, this deep data technology integration will enable Dynamics 365 customers to leverage all the features of Dayforce Payroll without the need for them to build or manage the integrations themselves. the vendor also launched a complete redesign of the Dayforce recruiting solution that focused on making the recruiter more productive by focusing on things like rules-based resume scoring. The vendor also offers native UK payroll and has enhanced its talent management capabilities with the release of compensation and learning management modules.

Infor

Global Human Resources is Infor's core HR solution for organizational planning, personnel system of record for employees and non-employees, benefits enrollment and life events, full absence management capabilities for plan definition and administration and a foundational set of self-service capabilities delivered on any type of device. The solution also has new-hire onboarding, offboarding, employee rewards management, employee relations, occupational health and employee safety. Tightly coupled with Global Human Resources (GHR) is a full talent management suite of applications built upon a framework that includes industry-specific content models as desired. This suite includes talent acquisition, compensation management, goal-setting, learning management, and advanced pre-hiring assessment. As with the other major players, Infor offers mobile functionality for iOS and Android OS. Infor continues to migrate its on-premise legacy customers to the cloud.

Infor's Workforce Management (WFM) solution covers labor cost controls, enhanced operating efficiencies and agility, time and attendance, scheduling, absence management and task management and also includes an industry specialization for healthcare scheduling. A COBOL-based, SaaS-delivered solution handles payroll and facilitates regulatory reporting, compliance with employee law, and the processing of payroll taxes. Other global regional packs deliver localized payroll capabilities. Infor offers cross-functionality with the vendor's financial management solution for WFM and HCM. The vendor competes with other vendors that offer enterprise resource planning (ERP) solutions, including Workday, Oracle, and SAP. For onboarding, Infor continues to differentiate itself by drawing on the vendor's suite and technology. This enables Infor to complement onboarding with unconventional ideas. For instance, the system can determine the retailers whose products might appeal to a new hire and then send him or her a related coupon ahead of his or her first day on the job. Infor offers Infor CloudSuite HCM Analytics, a new healthcare analytics solution designed to optimize workforce and operational effectiveness along with industry-specific content for predictive analytics.

Infor also combined two of its newest products, Infor Coleman, an artificial intelligence solution designed specifically for business users, and Infor Talent Science, a predictive behavioral analytics tool, to create Infor Team Dynamics. Infor Team Dynamics will extend the ability of Infor Talent Science to predict high potential talent based on not only they fit to the role, but also their fit to their new team and manager. Infor also announced the incorporation of Infor Coleman, into the company's Talent Science solutions. Coleman, a pervasive platform that operates below an application's surface, mines data and uses powerful machine learning to improve processes such as inventory management, transportation routing, and predictive maintenance, and now human capital management.

Kronos

Kronos offers a product line that includes its Workforce Ready solution acquired from SaaShr, and Workforce Central. Only Workforce Ready is on a public cloud. Workforce Central is designed for enterprise-sized users, but its usability is hampered because it is not on a public cloud. Kronos's 2017 partnership with Keysight Technologies Inc. and the resulting offering, NGA Time, was created partly to add functionality by coupling with this solution. Kronos Paragon, the vendor's implementation program methodology to bring legacy users to the cloud shows some positive movement by attempting to streamline Workforce Central's implementation.

In 2018, Kronos announced a number of additional features to Workforce Ready including Workforce Ready Employee Perspectives analyzes platform data to quantify employee attributes such as performance, reliability, and risk for better decision making. Also launched were Workforce Ready Succession Planning and Workforce Scheduler, as well as a new user interface with enhanced functionality will be available to all Workforce Ready customers starting this fall.

Oracle

Oracle provides a full suite of HCM solutions natively built on one platform. The suite encompasses the depth of HCM, including core HR, payroll, time and attendance, recruiting and onboarding, as well as talent management and workforce management. While Oracle continues to provide updates to the formerly named Taleo solutions for customers still on that platform, Oracle's sales strategy is to sell the entire HCM Cloud suite to new customers. The majority of net new customers comes from both large and mid-market sized companies. In addition, Oracle has launched a new program, Oracle Soar, to move its legacy on-premise customers to the Oracle HCM Cloud.

Oracle's Cloud Time and Labor product has matured significantly in the past year, leaving labor forecasting and optimized scheduling as the only functionality gap that remains. To address this, Oracle partners with best of breed vendors for each industry that requires this kind of advanced scheduling capability. Oracle has a very public partnership with Kronos for time and attendance. For retail and hospitality, Oracle partners with WorkForce Software and is in the process of securing partnerships with vendors who specialize in other industries such as healthcare and government emergency services. For field services and project management, Oracle Field Service Cloud also offers an advanced Shift Planning solution and Oracle Project Resource Management Cloud can be used to optimize resources to projects while leveraging worker data maintained by Oracle HCM Cloud and using web services to import schedules from those modules into Oracle HCM Cloud.

This past year, Oracle launched a free digital learning platform designed to help customers quickly and easily take advantage of the continuous innovation within Oracle Cloud Applications, called the Oracle Launch Pad. The platform provides comprehensive learning paths and task-focused modules with video tutorials and step-by-step instructions. The company is also enabling its ERP Cloud and HCM Cloud interfaces to support voice services such as Amazon Alexa. Oracle HCM Cloud mobile apps and chatbots also provide support for voice input and (device dependent) output/readback.

New customer support offerings are designed to help customers get more value from Oracle Cloud Applications including HCM Cloud, ERP Cloud, EPM Cloud, Supply Chain Cloud, Manufacturing Cloud and CX Cloud. Offerings include improved SLAs, 24/7 rapid response technical support, dedicated implementation support, proactive technical monitoring, success planning and adoption guidance, business process monitoring and guidance, Customer Success Portal, and a new digital experience platform for on-demand education resources. Oracle also offers a strategic partnership with IBM to provide Business Process Outsourcing for Human Resources delivered on the Oracle HCM Cloud platform. Together, IBM and Oracle will enable organizations to migrate to Oracle's HCM Cloud platform seamlessly.

Paychex

Paychex's HCM platform for SMBs, Paychex Flex, is a modular cloud solution that includes payroll, core HR, benefits administration, retirement administration, time and attendance, compliance, and integration with GL, analytics, and talent acquisition. Its analytics capability includes a "quick answers" feature that enables easy user access to the most frequently mentioned information in-context. Customizable dashboards enable users to easily view the results of analysis related to time and attendance, hiring, benefits, labor costs, and headcount changes. Paychex fosters ACA compliance by acting as a broker or agency with direct payroll integration. The solution emphasizes mobile usability with an intuitive ESS and UI.

This past year, the vendor announced Paychex Promise, a subscription-based service that protects clients against payroll interruptions. Its main feature is to extend the collection of payroll funds from an organization's bank account without service interruptions or insufficient fund charges. This feature enables business owners to continue paying employees regardless of disruptions in cash flow timing. Although Paychex offers good usability, it is not on par with vendors in the Leaders quadrant. We expect it to increase with the maturation of its analytics features. This, together with the eventual appearance of additional talent acquisition functionality, may enable Paychex to enter the Leader quadrant.

Paycom

Paycom's history is characterized by its expansion to provide functionality for WFM and HCM. Its features include payroll, time and attendance, payroll processing, benefits compliance, sourcing through onboarding for talent acquisition, performance management, compensation, planning, and learning. Because Paycom runs in a public cloud on a single application with a single database, the solution can process data in real-time. This benefits large employers and their efforts to comply with employment law.

Paycom's analytics are noteworthy for their focus on employment law and compliance, especially with the ACA and FLSA. The solution is also notable for its ability to allow employers to pre-screen candidates for tax credit-eligibility and secure them to help lower costs.

Although Paycom users have reported functionality and reporting limitations, ease of use, frequent automatic updates, and competitive pricing are positives of the solution. The vendor has recently expanded in size, with Paycom hiring for over 100 corporate positions, and announcing the opening of new offices in San Diego and Columbus Ohio.

Paycor

Paycor is a cloud-based suite spanning much of HCM. Paycor's capabilities include payroll processing, core HR, time and attendance, reporting, benefits administration, compliance reporting, analytics, talent acquisition, and learning management. Paycor has continued to hone its mobile functionality with updates to Paycor Mobile such as a mobile time clock, time off request and time-off balance, pay history, scheduling, and Spanish compatibility. The company acquired Newton Software in 2015 to enhance its talent acquisition functionality and has bolstered it with interview scorecards and a rating system for interviewees. Newton is now fully embedded into the Paycor suite and recruiting, and on-boarding is robust. Paycor also added Candidate Search, a new feature enabling users to leverage search filters to identify top talent. Paycor Learning Management was launched in October 2017.

Paycor announced Workforce Insights in April 2017, a feature that uses employee data from throughout the solution to provide interactive dashboards and customizable visuals for greater workforce insight. Partnering with Intuit's TurboTax in December 2016 enabled Paycor clients to view, print, and download W-2 information, allowing users to import tax information onto individual tax returns. The vendor has fully integrated 403(b) nonprofits into its Perform platform, making it an attractive option for companies within this sector.

Paycor primarily competes with the other "pay" vendors as well as ADP, as most of its clients employ fewer than 50 staff. That said, the vendor has begun to move upmarket and service companies with about 100 staff.

PeopleStrategy

PeopleStrategy provides a Cloud-based suite, eHCM, that provides a single end-to-end HCM solution specifically tailored to small and medium-sized employers. eHCM's capabilities include a full-service payroll module, benefits administration, time and attendance (with scheduling), talent acquisition, and performance management features. This solution also includes web-based mobile functionality, a reporting module with configurable dashboards and data visualization capabilities, and the ability to compare health plans side by side within its benefits administration module.

PeopleStrategy's solution differentiates itself with its focus on companies with 100 to 3,500 employees. eHCM does not require upgrades and fees when enhanced functionality becomes available. People Strategy's implementation program provides end users with constant communication when implementing eHCM and enables users only to deploy the features that they need. PeopleStrategy offers NavBar, a customizable navigation bar that provides a single access point to all information and tasks pertaining to a specific employee. The vendor has streamlined compensation features together with the ability to review, edit, and approve payroll. PeopleStrategy now provides interactive training content, such as pop up references and videos, as well as enhanced tax functionality that recommends state and local tax enrollments for work and home addresses.

Ramco Systems

Ramco Systems offers a multi-tenant cloud and mobile-based enterprise software in HCM and Global Payroll, ERP, and M&E MRO for Aviation. The company targets specific verticals including asset-centric organizations like aviation, logistics, equipment rentals; product & process-centric manufacturing; and people-centric staffing & professional services industries.

Ramco's payroll platform is configured to support around 45 countries and has partnerships to cover over 108 countries globally. Ramco provides the same product on-premise, on cloud, and as a hybrid option. The vendor also offers both public and private cloud option. In addition to cloud HCM solution, Ramco also offers a managed HR service which includes payroll, HR administration, employee helpdesks, and statutory lodgment services. Ramco's HCM solution includes modules and features to support the full employee lifecycle including Core HR, recruiting, talent management, benefits, time and attendance, compensation, and succession planning. The vendor also offers reporting and analytics across the various module of the HCM suite. Ramco offers both employee and manager self-service, covering core HR as well as performance management. Ramco offers ESS/MSS services in simplified Chinese, Vietnamese, Thai, Bahasa Indonesia, and Japanese languages, in addition to English, French, Spanish, and Arabic. The company also offers a hybrid mobile application accessible on Android, iOS, and Windows, and offers functionality for employee info, leave, expenses, time management, travel management, recruitment, and payroll.

With both EAM and ERP, the vendor is a theoretical competitor to Infor, Oracle, and SAP in deals where prospects want an enterprise-spanning suite. Ramco presents itself as a competitor to Workday, with core HR and integration with the general ledger. While the bulk of Ramco customers are internationally based, the vendor is planning to expand into the U.S. and U.K. markets, with fully compliant native payroll. This additional functionality, combined with its large international offerings, means Ramco is on track to become a full-throated, end-to-end option in HCM for global employers in the next few years.

SAP SuccessFactors

SAP SuccessFactors payroll and time and attendance functionality are still trailing behind the leading vendors in the HCM space. The vendor has committed to making improvements in this space, however. SuccessFactors highlights include SAP Jam, which enhances employee collaboration by being the social media platform for SuccessFactors. SuccessFactors Performance & Goals also offers trigger-based performance management, which enables employees and managers to request achievement feedback at any time and enables managers to track metrics. The SAP SuccessFactors Mobile iOS app enables users to access timesheets, time-off requests and approvals, search, organization chats, performance management, and reviews. Ongoing improvements to the existing Android OS app are being made to enable it to match this functionality.

SAP recently launched candidate relationship management capabilities as part of its SAP SuccessFactors Recruiting solution. Now with embedded candidate relationship management capabilities, recruiters can more efficiently manage the application and hiring process. Recruiters can manage candidate engagement end to end on a self-service basis across a multitude of channels. Also offered is SAP SuccessFactors Visa and Permits Management, as well as new and updated features to the SAP SuccessFactors HCM Suite around General Data Protection Regulation (GDPR) rules. The vendor also offers Upgrade2Success, a program that helps customers with on-premise SAP ERP Human Capital Management (HCM) solutions transition and expand into the cloud.

Snag

Snag, recently renamed from SnagAJob, has completely integrated PeopleMatter into its offering. Snag now offers a single integrated solution for WFM, recruiting, onboarding, learning, and performance management with a connect to hourly job candidates and hiring organizations interested in them.

Snag.work connects employers with hourly employees in the retail and restaurant space. Snag recruits and provides the background check of the full roster of "snaggers" and even conducts interviews. The company then connects these employees with employers looking for workers to pick up shifts. Snag then pays the employee right away, carrying the float. Currently available in Richmond and Washington, D.C., the company is looking to expand into other markets. In July, Snag announced that it will now offer its recruiting services to employers in Canada, the companies first international market. Canadian companies can now use Snag's innovative platform to create and post job openings and to manage the hiring process from start to finish.

SumTotal Systems

SumTotal Systems (part of Skillsoft) is one of the few HCM platform vendors that offers the full breadth of HCM solutions. SumTotal incorporates all key HCM components—core HR, talent acquisition, learning management, talent management, payroll and workforce management. SumTotal continuously invests in platform innovation leading to strong usability and functionality. The platform integration also allows for the unification of disjointed business processes such as the validation of certifications and training obtained in learning management when scheduling employees in workforce management. Employees that do not have the required training for a job they are scheduled, are identified to supervisors as part of the scheduling process, thus mitigating compliance risk and possible workplace accidents. SumTotal's learning management is the first to fully enable content aggregation across xAPI, CMI5, third party and custom content as well as unified access to Skillsoft's large corporate library. SumTotal integrates Skillsoft's multi-modal content, enabling organizations to develop talent through an employee lifecycle.

In addition, SumTotal offers three deployment types—SaaS-delivered via multi-tenant cloud, privately hosted in SumTotal, or on-premise. But customers always have the option of the latest version, regardless of deployment type. This allows for mixed deployments within the same organization, allowing for companies to account for specific union or regulatory driven deployment requirements while allowing other areas to take benefit of the cloud.

SumTotal offers a Netflix style design with a personalized experience covering a user's learning and development requirements as well as new gamification and social enhancements leverage data and analytics within the SumTotal platform to drive employee engagement. New talent acquisition functionality includes easy resume upload and AI matching to open positions while improvements to onboarding help new hires match with the appropriate mentors and map out their career paths.

SyncHR

SyncHR is a multitenant cloud solution designed around a single, inseparable core made up of HR, benefits administration and payroll. A second, patented time-based layer is applied to track and manage system transaction across time. This allows users to transact intuitively and accurately within the system across any timeframe—past, present or future. For example, users can make payroll corrections in the appropriate payroll period or view organizational charts from various time periods. Another patented aspect of this solution is its ability to tie HCM workflow to roles instead of employees, reducing the lost productivity employers otherwise face when assigning newly hired or promoted employees to the existing automated tasks of predecessors.

SyncHR's single dataset patenting has led to a great amount of usability and flexibility regarding analytics use potential. The vendor has partnered with Kronos's Workforce Ready solution for time and attendance functionality, together with a series of other vendors for talent acquisition. Additionally, SyncHR's platform includes a modern cloud extensibility layer, including APIs and MuleSoft, allowing users to leverage best-of-breed third-party solutions or easily connect to other corporate systems. This flexibility is particularly attractive to enterprises that want best-of-breed solutions or prefer an enterprise level solution with successes on core HCM. While this flexibility is attractive to some end users, the lack of native time and attendance and talent acquisition functionality is where SyncHR falls short compared to Ultimate Software, Paycom, and Namely. The eventual fusion of these features with its single dataset will enhance its competitive viability versus its competition.

In May 2018, SyncHR announced a partnership with myHRcouncil to enable clients to keep up with changing federal and state regulations pertaining to employment law and compliance. HR and finance users will benefit from its ability to provide legal information on Federal and 50 state employment laws.

Ultimate Software

Ultimate Software's UltiPro is a full-suite cloud-based HCM solution. UltiPro covers core HR, payroll, benefits management, the remaining breadth of WFM, recruiting, onboarding, career and succession planning, performance management, learning, compensation management and salary planning. Included in core HR are advanced reporting capabilities, predictive analytics, and a metadata-driven configuration platform and rules engine.

Ultimate continues to offer strong solutions for complex payroll, predictive analytics, recruiting, onboarding, and analytics. Ultimate users can use predictive analytics for both retention and performance, helping to lower turnover and increase productivity. Ultimate also offers prescriptive analytics via Leadership Actions and UltiPro Perception leveraging AI and machine learning to recommend specific actions for managers to recognize, engage, and develop the potential of their employees. In the past year, the company has made the following announcements:

- In July, Ultimate entered into a binding Letter of Intent (LOI) with respect to the acquisition of PeopleDoc by Ultimate Software. PeopleDoc is based in Paris, France, and has more than 1,000 customers with users in 180 countries. The addition of the PeopleDoc HR Service Delivery platform will offer new, person-centric features, such as an online employee help center and knowledgebase, HR case management, and employee file management.
- The vendor also launched Xander. Xander is an AI foundation that leverages Natural Language Processing (NLP) and advanced machine learning technology. Xander is capable of understanding structured and unstructured data and sentiment from open-ended text feedback and recognize over 100 different emotions and detect 140 workplace themes. Xander is embedded in UltiPro and the engine driving UltiPro Perception. Xander continuously learns and "gets smarter" over time, continually testing the relevancy and effectiveness of its predictions and suggestions.
- The vendor also offers UltiPro Connect, Ultimate's centralized integration hub designed to simplify and standardize integrations. In 2018, Ultimate extended options for customers with the most complex benefit needs with a premium offering, UltiPro Benefits Prime. In addition to handling everything related to dependent eligibility,

enrollment, and administration, UltiPro Benefits Prime offers a consumer shopping experience for enrollees, including embedded decision support (i.e., education tools—video, plan comparisons, and guided recommendations). It also provides tools such as EOI management and automation of benefits billing and reconciliation.

- UltiPro Perception users now have access to Mercer-Sirota's established surveys, question banks, and global benchmarks, all embedded within the UltiPro Perception solution. While Prescriptive Leadership Actions, added in UltiPro, reinforces practices by managers, who are guided to recognize, engage, and develop the potential of the individual.

 ## ROI Case Study: Ultimate Software

Swap.com · ROI: 355%

Swap.com deployed UltiPro to improve recruitment and retention and automate its processes. Having outgrown its old Paychex system, Swap.com was looking for a solution that would better meet its growing needs. With Ultimate Software, the company was able to increase productivity in hiring, onboarding, and benefits administration, while also using reporting and analytics to significantly reducing turnover.

THE COMPANY
Swap.com is an online consignment store offering pre-owned baby, kid's, maternity, men's and women's apparel and accessories. Swap.com is headquartered in Bolingbrook, Illinois, and has offices in Chicago and Helsinki, Finland. Sellers send their items to the Swap.com fulfillment center, where Swap.com checks the items for quality, individually packages the items and holds the items in its warehouse for sale on the online platform.

KEY BENEFITS
Swap.com was able to both automate and improve its

internal and external HR processes. Key benefits included: Reduced legacy software cost. By switching to Ultimate, Swap.com was able to pay less for its HR solution while at the same time receiving more functionality.

CUMULATIVE NET BENEFIT

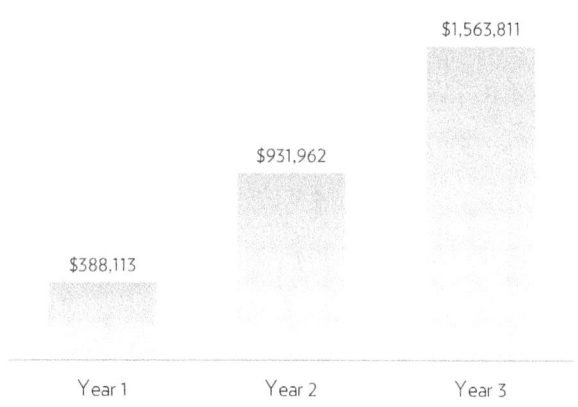

$1,563,811

$931,962

$388,113

Year 1   Year 2   Year 3

- **Increased user productivity.** The automation and overall improvements in efficiency to both hiring and onboarding, as well as benefits administration, increased the productivity of the HR team. These productivity savings represented anywhere from 15-25 hours a week and resulted in 15 percent of the project's benefits. Productivity increases in hiring and onboarding can be attributed to the fact that the HR team was able to pull information from an applicant's job board application. Applicants' information previously had to be reentered on the Swap.com website. Additionally, the employer was able to use UltiPro's analytics to track where it was getting its best applicants, reducing the number of job boards used from eight to three. The productivity in benefits administration was a result of automating the benefits process, meaning HR staff no longer had to manually enter this information.
- **Reduced turnover.** Using the new reporting and analytics offered by Ultimate, Swap.com was able to lower

their turnover rate from over 150 percent to approximately 100 percent. With the average cost of a new hire being $3,500 per employee, this reduction resulted in significant indirect savings that accounted for 60 percent of the project's benefits. Swap.com was able to achieve this by using UltiPro's reporting and analytics capabilities, which enabled management to better and more simply understand the factors driving turnover. By understanding the factors behind it, managers were better able to address turnover.

• **Error reduction.** Swap.com was able to eliminate errors in the manual entering of benefits. This had previously cost the company several thousands of dollars per year, but has not been the source of such lost productivity since the employer switched to UltiPro.

BEST PRACTICES

One of the most important things the Swap.com team noted was that prospective end-users need to be prepared for an implementation to become a major part of their work on a daily basis. This deployment also demonstrates the importance of developing and sticking to strict deadlines in order to time a new solution's go live to coincide flawlessly with the removal of the old one. Setting and keeping to these deadlines helped the team at Swap.com prepare for the roll-out while also continuing with their day-to-day activities.

· · · · ·

Workday

Workday delivers financial management, human capital management, and analytics applications. Associated products include learning, payroll, planning, recruiting, time tracking, benefits administration, talent management, absence management, compensation, and succession planning. Highlights include:

- Workday's learning functionality leverages the industry leading trigger-based model. This application enables employees to create, share, and consume content through browsers or mobile devices and has the ability to recommend content based on preferences and interests, peer popularity, and Workday transactions.
- Employers can require blended learning, instructor-led training, or external content at specific intervals, with reporting enabling them to understand the impact of their learning programs on key HCM measures.
- Workday Planning for budgeting and forecasting aims to unify financial and workforce planning through HCM and financial management collaboration.

Similar to competitors such as SAP SuccessFactors, Workday faces the challenge of enhancing payroll and time and attendance functionality. We do not recommend Workday for employers that need greater functionality for time and attendance, but this can be somewhat mitigated by partnerships. In the past year, Workday announced the acquisition of Adaptive Insights to create embedded analytics in a suite of applications for finance and HR. Since January, Workday has also acquired SkipFlag, Rallyteam, and Stories. bi. Workday also partners with Duo Security, a cybersecurity company that specializes in trusted access and multi-factor authentication (MFA) technologies. Workday also offers Workday Data-as-a-Service (DaaS), a cloud service that provides valuable data to customers to enable more informed decision-making.

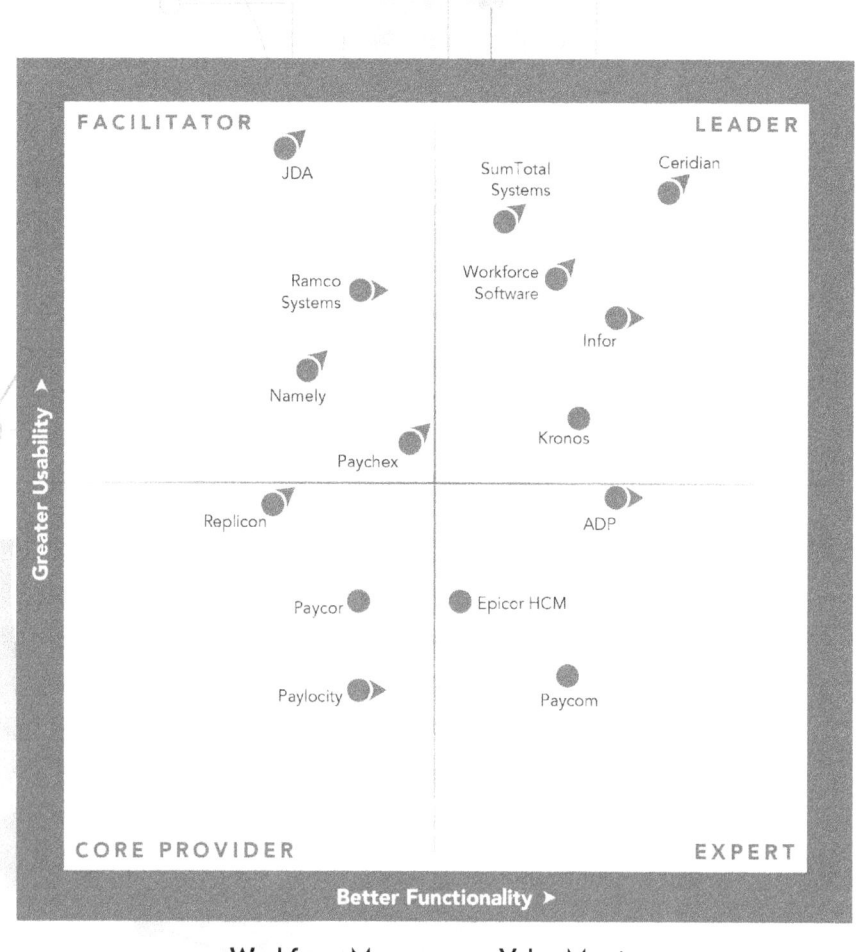

Workforce Management Value Matrix

Dot indicates current position. Arrow indicates trend in 2019 relative to others in the market.

CHAPTER 4
WORKFORCE MANAGEMENT

This year's Workforce Management (WFM) Value Matrix reveals that legacy players Kronos and ADP have little brand equity left to mask their users' mounting dissatisfaction. Against this backdrop, Ceridian has gone public with the best combination of functionality and usability found in this report, and SumTotal Systems and Infor have advanced farther into the Leader quadrant. As JDA, WorkForce Software, and new entrant Replicon prove specialization remains inescapable in WFM, Infor and new entrant Ramco Systems have identified the epicenter of human capital management (HCM) as the fulcrum to unify the enterprise.

WFM is the epicenter of HCM, where functionality for the essentials of employing people resides: payroll, core HR, time and attendance, scheduling, and benefits administration (Nucleus Research r97— *Value in HCM ripples from the epicenter outward*, May 2017). This is arguably the most important area of the enterprise to get right. The consequences of choosing a suboptimal vendor here are potentially paralyzing. At the same time, the risk of doing so is great: Some of the most storied brands in this space are today far from the best bets for workable WFM technology.

This year, vendors with superior technology for WFM have finally put considerable distance behind themselves and the promises behind trailing legacy competitors' brand equity—such that it is. Presenting themselves as the go-to choices for WFM, ADP and Kronos have for years survived off the fruits of brand equity that has gone stale. These fruits have left these vendors with a now difficult-to-manage, sprawling installation base deployed on an alphabet soup of products. Some of these users are on the new products, whereas others are on greenscreen or other ancient solutions.

· · · · ·

ADP

This year ADP has dropped just below the horizontal axis to enter the Expert quadrant. Complaints from users current and former have become loud enough to outweigh any favor the vendor may have earned through sheer years' worth of being While ADP is the first name that comes to mind when those in the industry think of payroll. Nucleus's analysis finds many ADP customers wish they had something different in place. The dissatisfaction appears to be a general frustration attributable to no one thing.

One challenge the vendor faces is that a large contingent of its existing installation base is on an older solution from ADP. Furthermore, in deals where ADP reaches these customers to present them with one of its new solutions for midsize or enterprise-grade employers — Workforce Now or Vantage HCM — users often choose the competition for reasons such as price and flexibility in the deployment.

ADP still uses a mainframe for payroll. This mainframe for payroll is highly sophisticated. ADP's payroll calculation engine processes payroll for an exceedingly large number of American workers via this IBM system Z mainframe. A multitenant application, it hosts and provides access to employers via the internet the same way a bank or financial institution also runs on a mainframe, yet acts like a cloud-based application. One area where ADP excels is in the functionality and usability of its mobile application, which sits atop proprietary assets of Kronos and ADP. Also, as noted last year, ADP is among the very best choices for global payroll.

Ceridian

With Dayforce, Ceridian provides full-breadth WFM on a single application producing a single dataset governed by a single rules engine. The vendor continues to reinforce and innovate functionality. In WFM, however, Dayforce is particularly mature. Global payroll is particularly strong for Ceridian, were it is able to compete against the likes of SAP, Ramco, and ADP.

The vendor continues to transform itself into a global player in WFM in other ways, too. In the most recent 18 months, Dayforce has added notable functionality designed specifically to facilitate compliance with the regulatory framework for employment in EMEA and Europe.

Nucleus has analyzed the ROI of numerous Dayforce deployments and finds robust results most recently in the grocery and restaurant industries. Users here and from other deployments of Dayforce express satisfaction with the intuitiveness of core HR records and report productivity gains attributable to the ease of use and mobility of ESS and MSS.

Epicor HCM

Cloud-based WFM functionality is available as part of a suite that integrates with Epicor's other solutions. WFM functionality covers payroll, benefits administration, time and attendance, absence management, and scheduling. The vendor also offers predictive analytics to identify staffing needs and provide insight into things like staffing demand and trends.

Known mainly as an ERP vendor, Epicor has an WFM solution that is part of Epicor HCM, whose installation base is primarily in the manufacturing, distribution, retail, and services industries. Users may link WFM or HCM with their other Epicor applications. They may also use Epicor HCM to manage their compliance with employment law. The solution offers both ESS and MSS.

Users note that the vendor's scheduling system, while fine for straightforward scheduling scenarios, struggles when faced with more complex needs. As is the usual in multivendor deployments, the time and attendance component of the Epicor HCM suite will feed information into another payroll application, if the employer uses something else.

Infor

For WFM, Infor relies on the strength of a growing collection of cloud solutions tailored to industry verticals. For example, for health-focused organizations the Health CloudSuite combines functionality for WFM with supply chain management (SCM) and financials. Here, SCM finds a close sister in Infor's suite for enterprise asset management (EAM). The CloudSuite strategy at Infor underscores the vendor's ability to meet users' needs across the enterprise, not just in WFM or HCM.

Coleman draws on natural language processing (NLP), image recognition, and machine learning to work in concert with the cloud-based analytics platform the vendor acquired in 2017, Birst. These predictions and recommendations reflect data not only in WFM, but elsewhere across the Infor enterprise applications and from outside the system, too.

Nucleus's analysis of Infor's customers in WFM found that users see marked increase in productivity. Users alsp express satisfaction with the scheduling solution, functionality for ESS and MSS, the mobile-friendliness, and processing for E-Verify and I-9 forms. It is notable that much of Infor's new business in WFM comes from disaffected former Kronos users.

JDA

The JDA solution is tailored for the needs of retailers by focusing tightly on their complex needs in scheduling and associated time and attendance. Nucleus's analysis continues to find that users in retail experience a transformation in their ability to predict, anticipate, and meet labor needs in a way that approaches precision.

The solution is likely the best for retail environments, where Nucleus's analysis finds JDA winning in head-to-head deals against Kronos and even Ceridian Dayforce and WorkForce Software when employees' time on the floor or elsewhere on the job must reflect the constantly changing variables in the omnichannel—e.g., sales history, foot traffic, shipments, customer behavior, and activity in ecommerce and in the warehouse and elsewhere.

The system enables employers to set fixed shifts for employees who want it and finds the best work for that person to do during these fixed times. Functionality can also be used for long-range planning, taking historical and real-time data into account. Scheduling also has an optimizer which accounts for employee attributes, including professional certifications and more, to help ensure that the employer has the properly qualified people on the job and to disallow the scheduling of minors when they may not work according to employment law.

For payroll, JDA will integrate with an employer's existing solution. ESS, a mobile-based feature, enables employees to swap shifts and managers to communicate available shifts to appropriate members of the workforce on short notice.

Dashboards give users side-by-side comparisons of locations that have seen recent overstaffing and enable them to adjust these schedules as needed.

The Store Optimizer found in JDA's solution helps to ensure that the right employee receives the right task at the right time and, moreover, assigns tasks to available staff with the appropriate skill sets based on the staffing schedule created. Elsewhere, users say the UI is intuitive for store workers to put in a time off request and check their schedules on mobile devices.

Kronos

Kronos remains in the Leader quadrant. Last year, Nucleus's analysis of user cases and aspects of Kronos's premiere offerings, Workforce Central (for enterprise-class customers) and Workforce Ready (for SMBs), suggested strongly that the vendor was faltering compared to the competition particularly as it had for a long time lacked a cloud offering. To rectify this, the vendor launch of Workforce Dimensions, a bona fide SaaS-delivered solution for WFM residing on Google's multitenant cloud architecture.

In the area of clinical scheduling, Kronos announced in Q1 of 2018 that it has entered a partnership with Spiral Software, home to AMiON physician scheduling software. This partnership also makes sense. Kronos's partnership here will make the vendor's offering more inviting to prospects in the healthcare industry. In addition, over the past 12 months Kronos has announced a major upgrade to mobile functionality, as well as enhancements to the vendor's ability to provide users with analytics. Despite these investments, we continue to here strong negative feedback around Kronos, particularly in regards to its (lack of) customer service.

Namely

The vendor's solution can handle users who employ up to approximately 3,000 in staff; employers in its installation base average 200 in staff, and the vendor says the solution's optimal customers employ anywhere from 20 to 2,000 in staff. Nucleus's analysis finds that the majority of Namely's new business wins are employers retiring a solution from ADP or Paychex, both of which are competitors for net new deals.

Namely now offers a managed version of its HR technology platform called Managed Services. This includes core functions such as payroll, benefits administration, and compliance, along with features such as time keeping. Customers have dedicated account managers who can do everything from run payroll to track benefits. Also offered is Namely Analytics includes detailed reports such as salaries, job changes, and attrition.

For WFM, the Namely suite encompasses payroll, core HR, time and attendance, and benefits administration. Last year the vendor introduced time management and, recently, expanded this to mobile for employees and managers. Updates to the iPhone application continue, and the solution is now iPhone X-compatible.

Users noted the modern look and feel of the UI, which offers a newsfeed resembling those found in consumer-grade social media. Additionally, employees can see everything in a calendar view whose look resembles popular scheduling tools and synchronizes with Microsoft Outlook and Google Calendar. For reporting, the vendor offers enterprise-grade reporting to small business users, with more than 80 standard reports available for use.

Paychex

For WFM Paychex Flex encompasses benefits administration, time and attendance, compliance, core HR and, of course, payroll. Every user has a dedicated support specialist who is available 24 hours per day every day of the year, including weekends and all holidays. Nucleus's analysis finds users noting usable dashboards and a streamlined integration with key applications, notably the GL. Paychex can handle time and attendance for organizations employing up to 10,000 in staff. This capability includes compliance with employment law. Advanced scheduling and budgeting, as well as analytics, are available too. Paychex has also launched an iris-scanning time clock to help employers eliminate buddy-punching. Paychex also has an intuitive UI for mobile, where ESS enables users to see and manage essentials of employment such as pay and rosters.

Paychex acts as a broker or agency, one of the top 25 in the industry. Moreover, the vendor has been a professional employer organization (PEO) for years. Paychex will engage with good-risk businesses to help reduce costs for various benefits. Paychex also offers functionality around new hires. As a customer's new hires come on board, Paychex immediately cross-references the pertinent information to see whether any tax credits are available to the hiring organization. Occasionally, these tax savings can be significant. The Paychex General Ledger Service (GLS) also integrates with Sage Intacct. The development synchronizes payroll and accounting data, and users may post payroll entries to Sage Intacct.

Paycom

An option for SMBs, The Paycom architecture offers a single application producing a single database rooted in the employer's origins in payroll. In early May 2018 the vendor revamped functionality for ESS. Improvements span both the desktop and mobile domains of Paycom's application. The UI has been redesigned, and new functionality centers on providing more immediacy to employees (e.g., shift clock-in and clock-out, expense report management, timesheet manipulation, etc.). As Nucleus's analysis has shown in general, a workable, modern portal for ESS translates to gains in productivity.

Paycom remains a strong option for SMBs. While lacking the depth that some of the enterprise-grade WFM vendors in the Leader quadrant have, Paycom offers a competitive solution that meets the typical SMB's fundamental business needs. Paycom users continue to report limitations in functionality and reporting, but positives include ease of use, dedicated account managers, frequent automatic updates, and competitive pricing.

Paycor

Paycor is a cloud-based suite for WFM that covers payroll processing, core HR, time and attendance, reporting, benefits administration, and compliance reporting. Since Q4 of 2017 the vendor has entered into two partnerships to enhance its abilities where benefits administration and payroll processing intersect — one with Employee Navigator, the other with Web Benefits Design. Paycor is also available in Spanish. Paycor also has 401K-related automation and integration.

Most Paycor customers employ approximately 50 in staff, and many of Paycor's customers are SMBs implementing their first-ever technology for WFM. Nucleus's analysis suggests ADP RUN, Zenefits, Namely, and others are among these competitors whether Paycor is facing or replacing them.

The vendor's mobile functionality helps employees with actions such as clocking in and out of their shifts, looking up their work schedules, viewing action balances, requesting time off, and viewing their paychecks and pay history. Recently, Paycor released capabilities in mobile to help users more easily navigate and understand their organizational structure. Paycor is particularly well-suited to several niches and industry verticals, especially nonprofits and small healthcare-related organizations. As we have noted previously, Paycor is a popular choice among smaller franchisees.

Paylocity

Paylocity is mostly for medium- and enterprise-sized users. WFM includes payroll, core HR, benefits administration, and time and labor. Payroll has data integration, which helps ensure regulatory compliance. Additionally, Paylocity is a registered reporting agent for the IRS and can provide users with a complete tax filing services.

Paylocity offers MSS and ESS, as well as more than 100 standardized reports with associated insight charts, as well as simplification for the on-boarding process to reduce time associated with new hires. Paylocity also provides a year-end dashboard for HR to engage in tax reporting or prepare W-2s. There is extensive automation in Paylocity of activities around expense filing, reporting, and reimbursement, saving both time and reducing errors related to manual entry.

Ramco

Ramco Systems is a cloud-based, SaaS-delivered solution for HCM, EAM, and ERP. The vendor also provides software for logistics management and for the aviation industry. WFM-related components found in the solution are core HR, global payroll, and time and attendance. ESS and MSS thread throughout the suite. With EAM and ERP offerings, for example, the vendor is a theoretical competitor to Infor, Oracle, and SAP (with or without SuccessFactors) in deals where prospects want an enterprise-spanning suite that includes WFM.

The vendor provides global payroll in 43 countries, which puts Ramco on similar footing with ADP and SAP in global payroll coverage. It is important to note that, despite this, Ramco works with Ceridian, Paylocity, and Paychex et al. to provide payroll in the United States and United Kingdom. Ramco brings credible competition to Workday with core HR and integration with the GL. The vendor also provides analytics in the form of Ramco Insights. Much of the functionality here centers on predictive capabilities around employee attrition, though customers whose experience Nucleus has analyzed point to the vendor's powerful analytics as the decisive factor in persuading them to choose Ramco. As for mobile, the native application is available for Android, iOS, and Windows.

Replicon

Replicon has a focus on time tracking and management and gross global payroll calculation. The focus is deep, and Replicon's product roadmap shows the vendor is playing a long game in applying the WFM-centric idea of time management across the enterprise—professional services management (PSM) is one example

The vendor's solutions comprise cloud-based functionality and applications for automating time tracking related to employees, projects, expenses, and resource scheduling. With all this, the idea is to track contractors' billable hours or employees' time, for example, against project costs and the like. Tight integration with the GL is, therefore, part-and-parcel of Replicon's solution. This helps ensure that all time-related data is accurate for accounts payable. The accuracy of time tracking functionality found in Replicon also facilitates compliance and informs complex scheduling, as well as predictive and prescriptive analytics, pay rules, and time off management.

Replicon also offers functionality to automate global time tracking and the calculation of gross pay, helping users with compliance via a mechanism to manage all time centrally. In the same timeframe, Replicon also offers an interface to facilitate users' ability to integrate with legacy on-premises ERP systems from SAP, Oracle, JD Edwards, and PeopleSoft.

SumTotal Systems

Our analysis shows SumTotal to be a wise choice for employers with complex needs in scheduling. Aside from the content found in Skillsoft, SumTotal has robust native capabilities in learning. These intertwine with WFM to support complex rules in scheduling. For example, a user may at once prompt an employee to renew his or her license to operate a forklift and disallow the scheduling of this employee till the system confirms renewal of this license.

Users of SumTotal may obtain the very latest version of the software SaaS-delivered via multitenant cloud, privately hosted by SumTotal, or on-premises. This is significant as it helps to minimize version sprawl by making it easy for users, no matter their circumstances, to ditch an old iteration should they wish. SumTotal encourages, but does not force, them to do so. This past year SumTotal introduced new dashboard widgets, several improvements to the user experience, and the ability for employees to trade on vacation and absences. The vendor's usability and functionality are improving faster than the average for the market

Users laud SumTotal's flexibility and nimbleness in handling complex scheduling and benefits management for wage employees. Our analysis of users' experience with SumTotal in fact finds the solution able to handle union rules that Kronos can't This is despite the latter's popularity, as Nucleus has noted elsewhere, as a legacy deployment that unions tend not to want to replace. In deals, moreover, users choose SumTotal over ADP citing the latter's lack of cohesion as a solution.

WorkForce Software

WorkForce covers time and attendance, staff scheduling, absence and leave management, labor analytics, and fatigue management, providing all new deployments in a modern cloud–based, SaaS-delivered platform. For payroll, WorkForce integrates readily with whatever the user has already deployed. Meanwhile, scheduling functionality from WorkForce is designed to handle exceptionally high levels of complexity, and for several years the vendor's go-to-market strategy has reflected this strength.

WorkForce has presently announced a new integration of its scheduling capabilities with time and attendance from Oracle. To promote and speed cooperation, WorkForce recently opened its platform to partners. Nucleus's analysis of these arrangements finds that in the case of SAP the go-to solution for these scenarios is WorkForce. Nucleus expects a similar outcome, in practice, for Oracle. Notably, WorkForce offers a comparableintegration for customers of Workday.

The ability to comply with the Americans with Disabilities Act (ADA) is now built directly into the solution. The vendor has also launched managed services for customers that would rather not have to deal directly with updating their system; instead, WorkForce will manage this and the system itself for them.

One of the strongest new features is the Android OS Mobile UI for crew management. A new mobile solution for crew management is presently available. It builds an entire timeline for a crew—down to what job each team member has done and for how long, meal breaks, etc. The new offering is for industries such as agriculture, energy or utilities, financial services (e.g., insurance), and construction, where fluid groups of field workers need a way to manage time with and without an Internet connection.

Talent Management Value Matrix

Dot indicates current position. Arrow indicates trend in 2019 relative to others in the market.

CHAPTER 5

TALENT MANAGEMENT

The talent management software market continues to winnow down, with only a handful of vendors surviving as standalone players against broader human capital management (HCM) suites with talent capabilities. Those that are still standalone will increasingly find their niche luster wearing off as customers look for an integrated view of talent that spans the entire employee lifecycle. We expect to see more partnerships and acquisitions in this space in the near future.

In this Value Matrix, we evaluated talent management vendors based on their product usability and functionality and the value that customers realize from the capabilities of the product (Nucleus Research, s142— *Understanding the Value Matrix*, September 2018). As a snapshot of the talent management solution market, this research is intended to help inform consumers about how well vendors are delivering value to customers, and what a prospect can expect moving forward based on the investment vendors are making today.

Leading vendors are innovating in the AI and NLP space, but adoption of advanced capabilities is still a challenge for many companies that are still focused on traditional backward-looking reporting and dashboards. The big challenge for talent management vendors (particularly the niche players) will be driving effective user adoption of the solutions that replace "gut feel" with actionable intelligence and insight.

The benefit of taking this approach is in shifting from a top-down to a bottom-up model of talent management, where the focus is on leveraging AI and automation to understand the engagement and skills of each individual employee, so managers can focus on not just what is communicated but how. A key part of this is the integration of collaboration and recommendations into HCM, particularly from a training and talent perspective, so insights can be captured in context and acted upon.

Other innovations on the horizon that will impact future leadership in the Matrix include:

- **Integrated coaching based on psychological profiles and communication styles.** Ceridian TeamRelate, for example, provides managers and employees with contextual coaching based on individual employees' values, convictions, and communication styles.
- **Bringing the "whole employee" into talent profiles.** With Oracle and others looking to integrate employees' skills learned outside the job — in volunteering efforts, for example — into their talent and performance lens, companies can have a more complete view of individuals' potential contributions.
- **Succession planning.** With all the leaders offering some capabilities for succession planning, it is increasingly becoming a necessary checkbox on the talent management RFPs.

· · · · ·

BambooHR

BambooHR is an option most viable for small- to medium-businesses. BambooHR's strongest component is its applicant tracking software (ATS), but also offers select features of HCM outside talent management. For BambooHR, functions are defined by role-based information, allowing for unhindered integration with the rest of the solution. BambooHR is most known for its strong applicant tracking system, which is supported by strong reporting functionality. This gives recruiters greater insight into how things like job descriptions are performing. In addition, customers noted its compliance capabilities and a satisfactory mobile application.

Cegid

Mainly based in Europe, Cegid is used in more than 75 countries across a diverse range of sectors including financial, manufacturing, government, health, education, and distribution. Cegid's talent management is focused into a single line of code, and services can be compartmentalized into stand-alone applications or as part of an integrated suite. These services include talent acquisition, onboarding, performance management, career and succession planning, compensation administration, learning and development, and Core HR. Cegid also offers Yourcegid Retail Y2 which is a modular, integrated platform for omni-channel retail strategies and powered by Microsoft's Azure platform.

Last November, the vendor announced the acquisition of Cylande, a leading provider of software solutions for retail. The combined company now supports over 70,000 stores in 70 countries globally. This year, Cegid announced that for its Talent Management (and other business units), Cegid is in partnership with Microsoft, whereby Cegid IT infrastructure will be gradually migrated onto MS Azure globally (Europe, North America and ultimately Asia). Additionally, in September, the vendor announced the acquisition of Loop Software, a web software company designed for accountants and EPTs/SMEs.

Ceridian

Ceridian offers talent management through its cloud HCM platform, Dayforce, and currently offers the full breadth of talent acquisition, onboarding, performance management, learning, compensation, flight risk, and most recently succession planning. The full talent suite is enhanced by embedded machine learning-driven analytics which is applied to the single set of data found in Dayforce. The vendor also offers Dayforce TeamRelate, a communication tool that not only identifies an individual employee's communication style and core convictions, but also provides managers with valuable insights into how to interact with that individual. In addition, TeamRelate offers real-time engagement tracking and provides managers with suggestions for practical, personalized coaching to improve their feedback and interactions. Dayforce provides comprehensive dashboards that present a full view of the employee as well as visual support graphics and tools. Ceridian also provides a Software Partner Program that makes Dayforce APIs open to third-party developers creating applications designed to complement and work in concert with Dayforce.

In October, Ceridian announced the availability of Dayforce Succession Planning. The new module will track key traits and succession information in an employee profile to help identify where employees are in their career path and allow organizations to create succession plans for any level position. In addition, it evaluates employees with a talent matrix that visually compares performance to potential, while using real-time data from across the HCM platform, such as performance history, credentials, compensation data, and flight risk analytics. In the spring, the vendor released two new talent management modules, compensation and learning. In addition to introducing new modules, Ceridian has redesigned Dayforce Recruiting to offer a new UI, rules-based candidate scoring, talent pools, configurable branding, and a mobile experience. Ceridian also rolled out enhanced predictive analytics capabilities that analyze key factors linked to flight risk.

Cornerstone OnDemand

Cornerstone OnDemand is a single application for all of talent management. Cornerstone OnDemand is for organizations employing 250 or more, but also offers Cornerstone Growth Edition for organizations under 250. The vendor offers several suites, including Cornerstone HR, Cornerstone Performance, Cornerstone Learning, and Cornerstone Recruiting. Last year, the vendor added major support for LXP for its learning delivery and now offers fully integrated capabilities in machine learning and predictive analytics. The system's machine learning continually analyzes employees' professional profiles and learning activity so that it may deliver content that is as relevant as possible to any given employee at any given time. Like other major vendors, Cornerstone also offers an implementation program to shorten the implementation cycle and help end users achieve fast time-to-value. Predictive analytics thread throughout all the new suites recommending courses, analyzing data and subscribing employees to playlists of learning content.

In the past year the vendor acquired Workpop, a recruiting space company that focuses on frontline employees. Workpop was acquired from the Cornerstone incubation space. Cornerstone also announced Cornerstone Frontline, a recruiting solution for companies hiring entry-level and frontline employees. Cornerstone also announced the acquisition of Workpop, Inc., which provides web and mobile solution for candidates and hiring managers in service-based industries.

Epicor HCM

Epicor HCM specializes in organizations with users ranging from 100 to 5,000 employees, Functionality for talent management included recruitment, onboarding, performance management, training, and employee development. The package can be used as a single-instance hosted via SaaS or on-premise. Epicor combined Epicor HCM with its flagship Epicor ERP solution in 2016, allowing manufacturers who use Epicor services to add functionalities including employee leave and actual time worked. This move was a part of Epicor's industry-specific strategy. we has found users who were dissatisfied with the solution, as well as others who had more positive reviews. The UI launched in 2017 has been received positively by end users.

Greenhouse Software

Greenhouse Software is best situated to serve companies which are medium-sized or smaller. As part of its talent management offering, the vendor has a candidate relationship management system as well as a predictive analytics framework that prioritizes talent acquisition. In addition to recruiting, onboarding, and inclusion, the vendor also offers a CRM solution. In the past year, the vendor has made upgrades to its onboarding, including adding pulled reports and automated tasks. In addition, the vendor announced the launch of new features to its reporting functionality, as well as an integration with IBM Watson Talent.

Haufe-Umantis

Haufe-Umantis offers performance management and goal setting, compensation planning, career development, learning, and succession planning. Haufe-Umantis competes with Workday, Oracle, SAP SuccessFactors, Cornerstone OnDemand, Lumesse, and Talentsoft across a variety of target markets. A German based company, the vendor acquired Chinese company Zhonghy Haufe in 2017 in order to further expand its professional development offerings in the Chinese market. We found usability mixed, with users noting that the suite was not user friendly.

HireVue

HireVue provides video interviewing software and assessments with Hiring Intelligence and its HireVue Video Interviewing platform, using a combination of industrial/organizational science and predictive artificial intelligence to shorten hiring time. HireVue is available worldwide in over 30 languages. Video interviews can be completed at the candidates' leisure, and then socialized amongst stakeholders. HireVue also offers game-based assessment tools that analyze verbal and non-verbal cues to help rank candidates for the display of important competencies for success in the job. In October, HireVue released Pre-Built Assessments, an update to the HireVue Assessments platform. Pre-Built Assessments are available for retail associates, sales representatives, customer service associates, and software developers. This past year, HireVue completed the acquisition of London-based MindX—a game-based candidate assessment company. HireVue plans to integrate MindX's advancements measuring cognitive ability with machine learning into its Assessments platform. In March, the vendor released new scheduling features for its Coordinate solution, which allows candidates to self-schedule and reschedule appointments, and gives recruiters more autonomy for managing events and interview lineups.

IBM Talent

IBM offers solutions including sourcing and recruiting, compensation and rewards, employee engagement, learning, employee assessments, onboarding, and performance management. The talent management service combines Kenexa-inspired functionality with IBM Watson's tools and power. The vendor also offers a large partner network for everything from assessments, screening, recruitment, job marketing, and more. In November, IBM announced a new partnership with Kronos, to bring its Watson AI platform to the Kronos's workforce management software suite, Workforce Dimensions. The IBM Watson Career Coach will interact with employees, via chat on their mobile devices, to offer career guidance, training courses, and milestones regarding a raise or a promotion.

iCIMS

iCIMS is a best-of-breed recruitment software provider. iCIMS enables organizations to manage and scale recruiting through a full product suite and network of more than 230 integrated third-party software solutions delivered within their platform-as-a-service (PaaS) framework, called UNIFi. iCIMS supports turnkey integrations with more than 1,000 HCM/ERPs including Workday, SAP, Oracle, Infor, Microsoft, Ultimate and ADP. iCIMS is particularly well-equipped for enterprise organizations scaling globally, along with companies with high-volume hiring.

Overall, customers reported higher satisfaction with customer service as compared to last year, and recent UI updates have improved usability. iCIMS is a partner in Google's initiative, Google for Jobs, to improve job searching and help candidates find best-fit jobs more efficiently. iCIMS is also an integration partner with Google Cloud Talent Solution, which enables employers to optimize their career site for the search engine, so candidates can easily find opportunities personalized to their specific interests including location, salary, responsibilities, experience and industry.

In January 2018, iCIMS acquired TextRecruit, the mobile candidate engagement platform that uses text messaging, live chat and artificial intelligence (AI) to hire people. In June, iCIMS announced a new partnership with Ultimate Software. The two companies now offer a seamless platform-to-platform integration of iCIMS' talent-acquisition solutions with Ultimate's suite of HR, payroll and talent management suite.This past year, the vendor also launched iCIMS Offer, an end-to-end solution that expedites the job offer and approval process as well as Prime Connector, a dedicated toolkit for employers to configure, administer, and run third-party talent acquisition applications, such as background screening, without technical support.

Infor

In terms of talent management, Infor's functionality covers a wide swath: talent acquisition; employee recognition; advanced pre-hiring assessment; and management for many more facets, including transitions, goals, compensation, succession, and learning. Infor also offers Coleman, its AI platform for CloudSuite applications designed for enterprise-grade businesses, as well as Brist, a cloud-based analytics platform. Infor's talent management functionality pulls many positive elements from the rest of the vendor's enterprise suite. For example, with onboarding, Infor's solution identifies retailers with products that could be enticing for newly hired individuals and sends them relevant coupons in advance of their first day of work. Infor caters to several micro-verticals and their unique needs by means of tailored CloudSuites. Examples range from healthcare and hospitality, to retail, manufacturing, and the public sector.

This year, the vendor added new functionality to Infor Talent Science by adding capabilities for identifying the "best fit" for a person, personalizing learning plans, suggesting the optimal career path for a person and identifying the makeup of a high performing team (called "team dynamics"). The upgrade of Infor Talent Science incorporates a new version of a machine-learning profile creation system to increase the precision and validity of the prediction models when organizations tap into the data of tens of millions of candidates on the global Infor Talent Science database. Additional announcements this year included the launch of new analytics capabilities to include inclusion & diversity metrics among other talent KPIs. Infor also launched optimized mobile. This extended existing mobile capabilities for gamification of recognition, profile updates for skills & competencies, completion of tasks and forms for transition management (transition management includes pre-hire, onboarding, cross-boarding and offboarding).

Lumesse

Lumesse, a European company, offers Empower, a solution very close to thoroughly covering functionality for talent management. The service is SaaS-delivered and can be cloud-based or hosted in a private environment and offers functionalities for succession planning, career pathing, development planning, and learning. Cloud-to-cloud connection between TalentLink and SilkRoad comprises the Onboarding capability. Lumesse also has a partnership for LinkedIn integration and is the only system whose integration is commercially available. TalentObjects is Lumesse's offering for talent acquisition and is designed for the Salesforce 1 Mobile Platform. TalentLink is geared for medium to large companies and their talent acquisition pursuits, while i-Grasp is for enterprise users.

Lumesse offers strong functionality for large global institutions that operate multiple brands. In October, Saba announced plans to acquire Lumesse, with the acquisition closing in November. In addition, the vendor released Lumesse Connect, an employee app that gives HR managers, executives and employees a common communication platform at all times, on all devices. In April, the vendor launched the Lumesse Talent Acquisition Marketplace a third-party marketplace for applications.

Oracle Talent Management Cloud

Oracle Talent Management Cloud functionality spans sourcing, recruiting, employee development, performance and goals management, career development, learning, and retention predictors and indicators. The past several years have seen Oracle expand its functionality. Additional new functionality includes mobile learning, features to support continuous performance including anytime feedback and check-ins, tools to shape and market the employer brand in sourcing, and enhanced capabilities to streamline and automate the recruitment process. Employees' public profiles enable them to share their interests and receive recognition and feedback, while learning continues to expand with collaborative approaches. Overall, recent functionality investment has focused on engaging passive talent, providing more candidate self-service abilities, eliminating barriers to internal mobility, and automating the recruitment process.

In October, Oracle announced the launch of its new artificial intelligence-powered innovations and other updates to Oracle Human Capital Management (HCM) Cloud. These features included scalable HR concierge with digital assistants, mobile responsive design, configurable action lists for represented worker processes, LinkedIn recruiting integrations, smart sourcing, self-learning risk management, and strategic workforce planning.

PageUp People

PageUp People functionality covers talent recruitment, onboarding, education, performance management, and succession planning. The vendor does best with contingent and contract labor. For learning, the vendor offers the Everyday Learning while the Everyday Performance App provides a mobile application that allows managers and employees to share their goals and receive feedback instantly. While most users were happy with the overall UI, problems were reported when using advanced configurations.

PeopleFluent

The PeopleFluent suite offers functionality for talent acquisition, onboarding, performance management, professional development, workforce analytics, learning management and content, succession planning, organizational planning, compensation planning, and workforce compliance. While the solution will accommodate users with employee populations as low as 2,000, most customers are in the enterprise space, typically having 20,000 plus employees. PeopleFluent's talent acquisition is designed for high volume scenarios, and the vendor typically finds itself most often facing off against Oracle, Workday, and SAP SuccessFactors. Customers that we spoke with cited talent acquisition as the number one reason that they went with the vendor.

In May PeopleFluent was acquired by Learning Technologies Group, which provides PeopleFluent with a global footprint through LTG's international presence. In October, the vendor announced that NetDimensions had joined PeopleFluent's talent management offerings. NetDimensions' best of breed LMS will now form a core part of a newly created learning suite. This past year, PeopleFluent announced a strategic partnership with StaffDNA, a talent solutions company that serves the healthcare workforce. Under the agreement, the organizations will deliver vendor management service (VMS) resources to clinicians and healthcare providers throughout North AmericaThe vendor also announced the launch of Talent Management Essentials. This is a guided implementational tool that streamlines configuration as well as integration of client data. The approach leverages pre-configured solutions to enable clients to incorporate best practices to quicken implementation and time to value.

Saba Software

Saba delivers talent development solutions that bring together capabilities across learning, performance, engagement, recruiting and workforce planning. With deep performance and program insights, Saba connects the success of the employees to the success of the business. Saba offers two platforms—Saba Cloud, designed for the medium to enterprise market segment, and Saba TalentSpace, designed for the small to medium market segment.

In November, Saba completed the acquisition of Lumesse, a leading provider of talent acquisition, talent management and learning experience technology in Europe. The acquisition gives Saba additional global reach, with in-region experts and deep local understanding in the largest markets around the world, while providing future-proof depth in the most pivotal areas of the people journey—from attracting, engaging and onboarding the right candidates, to developing and coaching people to the highest levels of performance, growth and engagement. In the past year, Saba has enhanced connectivity between learning and performance modules, as well as made improvements to its mobile app includes feedback, talent view, 1:1 meeting features and learning capabilities Saba also offers new rewards and recognition capabilities across both the learning and performance modules with the addition of an in-app rewards store, where employees can redeem points earned from learning, impressions and achievements for tangible rewards

There is also an expanded content integration framework, additional out-of-the-box content connectors, and new content partnerships as well as new integrations including LinkedIn Learning, Tango Card, UltiPro, and Job Target.

 ## ROI Case Study: Saba Software

Sunrise Senior Living · ROI: 1,043%

Shortcomings with the existing learning technology at Sunrise Senior Living complicated and slowed program administration, community management, and new-hire onboarding. In search of a replacement, the employer turned to Saba Software for integrated learning and performance management, deploying the latter to improve overall management of a leadership development program and boost its capacity to accept participants, as well as eliminate related manual workflow. The result was a solution that interfaced much better with other systems in place at Sunrise for human capital management (HCM) and improved the daily efficiency of staff at headquarters and across 280 locations in North America.

THE COMPANY
Founded in 1981, Sunrise Senior Living operates resident-centered communities for approximately 30,000 seniors in Canada, the UK, and the United States. These locations accommodate short- or long-term stays and feature independent living and the full breadth of senior care. Headquartered in McLean, Virginia, the company employs more than 26,000 across North America.

CUMULATIVE NET BENEFIT

$13,422,510

$8,642,028

$4,099,546

| Year 1 | Year 2 | Year 3 |

KEY BENEFIT AREAS

Following deployment of Saba, the employer experienced benefits in the forms of costs saved, gains in productivity, and other efficiencies. Following are details:

- **Software savings.** The first-year software as a service (SaaS) subscription fee for Saba cost less than the old platform's final year of licensing, maintenance, and support fees. Additionally, the outgoing vendor was charging extra fees for technical support and for tickets beyond the 200 allotted annually for base support. In the final 12 months on the old system, Sunrise Senior Living was careful to avoid these fees. Had the employer paid these, the difference in cost would have persisted and grown even after the purchase of an additional 2,000 yearly recurring seats starting in the second year of the Saba deployment.
- **Redeployed IT staff.** After the new cloud solution went live to replace the old, on-premises one, Sunrise Senior Living was able to redeploy one full-time IT staffer. This employee's annual fully loaded cost to the organization equaled 40 percent of total direct savings the employer realized with the new technology.
- **Productivity gains at headquarters.** Centrally located learning-focused staff at Sunrise Senior Living saved a

great deal of time across a broad range of activity. Some of this came with newfound automation in learning-related onboarding of new hires yielding gains in productivity anywhere from approximately 3 percent to 11 percent. Elsewhere, built-in, configurable logic and prescriptive rules in the Saba system were instrumental in eliminating manual workarounds that the old system used to exhibit whenever an existing employee moved from one location to work permanently at another. Headquarters-based learning development program managers saw an overall gain of 35 percent in their productivity.

- **Gains in productivity across North American communities.** Team leaders across 280 Sunrise Senior Living communities in North America are on the new system. This has yielded several million dollars' worth of gains in productivity. Each location employs a business office coordinator who manages the community's learning with the help of other department coordinators, also onsite at every location. With Saba in place, these leaders save time every day, at every location — with access to reports featuring far more detail than the old system's.

- **Greater alignment of learning activities.** In addition to the LMS, Sunrise Senior Living deployed functionality from Saba for performance management. The employer uses the latter unconventionally, as a way to drive participation and efficiency in the learning development program. Shortcomings in the old system forced the department to limit participation to 100 employees every spring and fall, but now there is no cap on the biannual roster. This was not even considered possible with the former LMS, whose process was manual.

BEST PRACTICES
Before looking for something new, Sunrise Senior Living distributed a survey to a subgroup of employees who were representative of the end-user base and incorporated

their input in the selection process. The partnership with IT proved central from selection to the implementation, helping the employer determine that deployment of a new learning management solution was the biggest need. Leadership agreed that any additional functionality that might come with the new system would be welcome, but secondary in importance to a more intuitive user experience and the absence of line-item expenses such as those the old vendor charged. Lastly, field operations confirmed for headquarters that staff training in Saba would be straightforward.

· · · · ·

SAP SuccessFactors

SAP SuccessFactors offers a full-service talent management solution, with particular strengths in learning and performance management. SAP SuccessFactors also offers mentoring capabilities as part of the SAP SuccessFactors Succession and Development solution, leveraging an intelligent algorithm to automate the matching of mentors with mentees. Learning continues to be a strong offering, with options for the learning management system (LMS) and the SAP SuccessFactors Learning Marketplace, which combines the LMS with SAP Hybris ecommerce, sales and marketing capabilities so that organizations can sell training to their external business ecosystem (dealers, franchises, partners, customers, etc.).

Customers continue to rate it highly and note it as a distinctive feature. Also offered are career site builder and social networking and collaboration capabilities, allowing employers to easily design, create and manage responsive career sites as well as create online communities for new hires. The mobile UI is particularly user friendly, with customers noting its simplicity and ease of use. Over the past year, SAP SuccessFactors announced the launch of candidate relationship management capabilities as part of its SAP SuccessFactors Recruiting solution. These capabilities enable companies to create and manage talent pools, and view candidates in a single profile, including all candidate engagement, activity, and history data. They also support e-mail campaigns, so recruiters can engage with talent pools and track campaign results.

SilkRoad

SilkRoad continues its focus on enterprise strategic onboarding as a tool to drive business outcomes. The SilkRoad Activate platform integrates technology with content and workflow to enable individualized new hire journeys, from offer delivery through year one engagement. They work with medium and larger enterprises to engage new hires early, accelerate time to value, and improve talent retention. Activate also includes rapid integration with other HR systems, analytic dashboards and reporting, and best practices consulting and advisory services.

The overall UI remains a problem for some users, but SilkRoad's onboarding remains one of the best in the business. SilkRoad has continued to invest in its onboarding solution, most recently in September announcing the expansion of its onboarding offerings. This expansion included SilkRoad Activate Offers, a solution that enables the creation, approval routing, digital signature, tracking and analysis of offers. Also announced was SilkRoad Activate M&A Onboarding, which covers the administrative and compliance tasks related to onboarding many new employees in a short timeframe following a corporate transaction. Lastly, the vendor added SilkRoad Activate Offboarding, a feature that automates critical offboarding tasks and coordinates activities across all stakeholders. We find that the organizations most likely to achieve high ROI are those with significant employee onboarding, engagement, and talent retention needs.

 ## ROI Case Study: SilkRoad

Accenture · ROI: 214%

To replace an outdated, primarily manual workflow for onboarding new hires, Accenture implemented and deployed SilkRoad Onboarding. Through automation, the organization eliminated faxing, scanning, printing, and mailing of compliance-driven and other documentation for hiring and reduced related labor for staff. These improvements saved labor expenditure for HR and reduced the time it previously took to bring new hires to full productivity.

THE COMPANY

Accenture provides professional services and solutions for strategy, consulting, digital, technology, and operations across more than 40 industries. The organization employs approximately 442,000 people and services clients in over 120 countries.

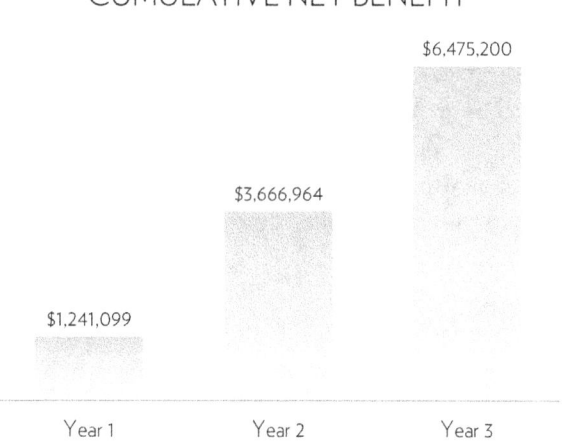

CUMULATIVE NET BENEFIT

$6,475,200

$3,666,964

$1,241,099

Year 1   Year 2   Year 3

KEY BENEFITS

The deployment of SilkRoad Onboarding brought several benefits. Some came in the form of eliminated or avoided costs. The rest were vast gains in productivity:

- **Reduction in hard copy documentation.** The deployment of SilkRoad Onboarding greatly reduced Accenture's paper-based administration necessary for the hiring of new employees. Some of this material was compliance-related. With automation and the digitization of this documentation, the organization saved significantly on associated printing, faxing, mailing, and scanning.
- **Redeployment of staff.** With the reduction of costs associated with hard copy documentation, plus elimination of data entry, administrative-related labor decreased. Because of this, Accenture was able to redeploy the equivalent of one full-time staff to other duties.

- **Compliance.** By switching to an automated, modern solution for onboarding, Accenture expected to avoid fines over the next 12 months. These would have been for falling out of compliance with hiring-related regulations.

- **Productivity gains.** Among HR and other staff responsible for onboarding new employees, time savings were significant. On average, these employees in aggregate saved 75 percent of their time in onboarding several thousand new hires in a year's time. Likewise, new hires now had a portal to visit where they were able to take care of paperwork and other necessities, such as connecting with new teammates ahead of the first day of employment. Ongoing, this has been a key factor in drastically reducing the time new hires need to become fully productive upon beginning employment.

BEST PRACTICES

In some companies the recruiting and onboarding departments collaborate to carry out onboarding. For other employers, including Accenture, these are mostly separate departments. Therefore, it is important for an organization to understand under which category it falls or, perhaps, rethink which of these approaches it wants to emulate moving forward. Notably, the technology for onboarding has evolved considerably since Accenture first deployed SilkRoad. Over the intervening years the employer has been able to make use of more progressive capabilities as these become available in the system. No tool for onboarding will meet 100 percent of any organization's needs, but the right solution will be flexible and do the best job of mitigating pain points such as those covered in this report.

· · · · ·

SmartRecruiters

SmartRecruiters offers functionality for recruiting software, solutions for recruitment marketing, a platform for collaborative hiring, and dashboards that help users manage their hiring campaigns. These hiring campaigns include a compliance module and enterprise analytics. SmartRecruiter currently offers thirty languages. Support for multiple contracts in LinkedIn Recruiter reduces administrative work while an internal job widget assists users with customizing their application experience and tracks those changes on corporate career portals and intranets. The Winter 2018 product update included custom data reports and integration of regional social networks such as DACH's Xing and China's WeChat.

SumTotal Systems

Skillsoft acquired SumTotal Systems in 2014 and in 2017 they succeeded in consolidating workforce management, talent acquisition and talent management onto one platform. Also in 2017, SumTotal enhanced its set of functionality included with all learning and talent deployments and now position what was formerly elixHR as SumTotal's Core Platform. The Core Platform now features in-memory data crunching for predictive and prescriptive analysis as well as a redesigned social and gamification platform along with enhanced reporting and analytics, and lifecycle career development functionality.

SumTotal's approach to social engagement allows its clients to choose their own third-party social media solutions (e.g. Yammer) or use the native social platform built into the SumTotal platform. With learning, SumTotal supports a learner experience platform that aggregates content from third parties and massive open online course (MOOCs). SumTotal is also the only LMS to support Percipio Experience Services, providing seamless integration with Skillsoft content include inline playback, support for curated channels and support for multiple learning modalities. SumTotal now leverages AI-driven processes to fill candidate pipelines within Talent Acquisition by auto-matching internal and external prospects to open job requisitions. This same type of job matching is utilized in Career Planning in that it allows employees to be matched to recommended career paths and then build development plans to meet short term and long term objectives.

The newest iteration of SumTotal offers a single code based solution available in multiple deployment options: software-as-a-solution, a private cloud solution, and a system installed directly on the premise, although ~85% of all new sales are for the multi-tenant SaaS offering. With the most recent software release, SumTotal now offers a SaaS Extended Deployment (SaaS EXT) solution that provides customers the opportunity to deploy on an automatically upgraded SaaS platform while having extended time for the testing and validation required in many regulated industries. In the past year, SumTotal announced the integration of HelloSign's platform for legal agreements with mobile-first, secure, compliant and legally binding eSignatures and forms into SumTotal Talent Acquisition, as well as strategic partnership with STRIVR's Virtual Reality (VR) training platform. SumTotal also implemented Percipio Experience Services (PES), Skillsoft's recently launched Content-as-a-Service offering. PES offers a set of APIs that enable micro-learning and multi-modal learning within any learning platform with SumTotal being the first LMS to support PES.

Talentsoft

Talentsoft offers functionality for recruitment, career planning, compensation management, competencies and performance management, and e-learning. It also offers embedded analytics in an innovative freemium business model. The vendor competes mainly in Europe, with a presence on all continents and offices in 15 countries. In Europe, Talentsoft's biggest main competitors are Oracle, SAP SuccessFactors, Cornerstone OnDemand and, increasingly, Workday. The company is currently looking to expand its presence into the North American market, where it already counts 50 customers in Canada. The vendor recently launched a core HR function, separating itself from the stand-alone talent players. This makes sense, as we view the best of breed application market as contracting, where only a few true standouts or niche players will survive. In the meantime, users report strong integrations with core HCM vendors.

In June, the vendor announced new tools including flexible performance appraisals, enriched, on-demand training programs, and embedded analytics in its solution. Talentsoft also announced a strategic partnership with payroll vendor Raet to provide a seamless end-to-end user experience.

FACILITATOR

LEADER

Greater Usability ►

Acumatica

Unit4

SYSPRO

SAP Business ByDesign

Oracle NetSuite

Microsoft Dynamics 365 for Finance and Operations

Qualiac

FinancialForce

Deltek

Infor CloudSuite

Microsoft Dynamics 365 Business Central

Oracle ERP Cloud

SAP 5/4HANA

IFS

Rootstock

Epicor

Sage

QAD

VAI

IQMS

Aptean

Plex

CORE PROVIDER

EXPERT

Better Functionality ►

Enterprise Resource Planning Value Matrix

Dot indicates current position. Arrow indicates trend in 2019 relative to others in the market.

CHAPTER 6
ENTERPRISE RESOURCE PLANNING

Enterprise resource planning (ERP) software operates as a central system of record for many organizations, facilitating day-to-day operations and tracking enterprise critical data. The solutions delivered by ERP vendors today often look to provide the visibility and control that enables more efficient operations and better decision-making. Regarding cloud adoption, many industry-verticals remain laggards because maintaining 100 percent uptime has become customers' primary concern.

In this Value Matrix, we evaluate ERP market vendors based on their product usability and functionality and the value that customers realize from the capabilities of the product. As a snapshot of the ERP landscape, this research is intended to help inform consumers about how vendors deliver value to customers, and what can be expected moving forward based on the investments vendors are making today.

Recently, the ERP market has made significant strides towards bringing intelligence to back-office processes. As cloud offerings mature, vendors are bringing functionality that exploits the flexibility and scalability of cloud platforms. Vendors have consistently made announcements and product releases around capabilities that will be delivered on their platforms and the extended services that these technologies will facilitate. Despite the cloud's slow maturation process, there are early signs that vendors will be able to differentiate based on these services and deliver new value to customers. Although the scalability of the cloud allows vendors serving small- and medium-sized businesses (SMBs) to meet the needs of large enterprises, new market segmentation is making solution identification easier for customers.

Verticalization and delivering more tailored functionality continue to be at the center of the value proposition for many vendors. Only a few vendors claim to have the functionality to service any business, and fewer still can deliver a solution platform flexible enough to enable customers and partners to build the capabilities needed in a cost-effective manner. As a result, customers operating in niche micro-verticals should be better able to determine if a vendor can serve their needs based on the specific business processes that the ERP can address.

Vendors are investing in a suite of technologies such as Internet of Things (IoT), artificial intelligence (AI), natural language processing, and bots. Since the switching costs are particularly high for ERP systems, ensuring that a new deployment will service the needs of the customer well into the future remains a vital consideration. Vendors are caught between addressing a customer's current business processes and demonstrating how they will continue to do so with their development roadmap as those businesses change. Although the number of concrete business cases for the above technologies is growing, the value delivered to the end customer is often still unrealized. We predict that vendors capable of demonstrating Industry 4.0 capabilities, will better differentiate themselves in the coming years.

· · · · ·

Acumatica

Acumatica is a Leader in the 2018 ERP Value Matrix. The vendor continues to lead the way in usability, serving several industry verticals, including commerce, manufacturing, and field service. Since that last Value Matrix, Acumatica has added a Construction Edition and Distribution Edition to its vertical offerings. With triple-digit growth, Acumatica has aggressively expanded its partner ecosystem to meet the needs of a diverse customer base, adding over 60 new value-added resellers (VARs) in the last year. As a born-in-the-cloud solution, Acumatica is delivered by subscription in the software as a service (SaaS) model.

Having recently secured Series C funding to bolster its investments in technologies such as AI and ML, Acumatica has stressed that it is working on practical applications that deliver value to customers. High usability and flexibility are at the center of Acumatica's development philosophy and the vendor is looking to automate processes that it already delivers, such as with multi-entity accounting when making payments to multiple locations from a central location. In its second update in 2018, due for release in September, Acumatica is bringing a host of incremental improvements to each of its vertical solutions focused on increasing customer satisfaction. For example, with its Manufacturing Edition, the vendor is in the process of delivering project manufacturing on top of its project accounting capabilities and advanced planning and scheduling. Additionally, mobile applications will continue to be a core focus of the second release of 2018, following on from the mobile dashboard capabilities that were included in the R1 release earlier this year. Predominately serving SMBs, Acumatica is continually looking for ways to improve its customers' user experience, often taking feedback directly from customers on what features should be delivered. This is reflected in the vendor's approach to Industry 4.0 technologies, which centers on fitting the technology to the business case rather than the reverse. The focus on usability and customer value should continue to serve the vendor well in future editions of the Value Matrix.

Aptean

Aptean is an Expert in the 2018 ERP Value Matrix, focusing on manufacturing, distribution, and logistics, as well as delivering significant industry expertise with its solutions. Aptean's vertical solutions cover a plethora of industries including food and beverage, process manufacturing, chemical, pharmaceuticals, retail, wholesale distribution, automotive, metals, discrete manufacturing, electronics, and medical device manufacturing. Supporting midmarket customers across 74 countries, Aptean complements its ERP capabilities with warehouse management systems (WMS), manufacturing execution systems (MES), and enterprise asset management (EAM) solutions.

Since the last Value Matrix, Aptean has made the strategic acquisition of IndustryBuilt, solidifying its position in food and beverage verticals. IndustryBuilt's core product, JustFood, is built on the Microsoft Dynamics 365 Business Central platform and is available on-premises and via the cloud, delivering end-to-end manufacturing, logistics, SCM, and warehousing functionality designed for the mid-market. The move by Aptean helps the vendor's coverage of some sub-verticals, complementing Aptean Ross ERP's functionality, and increases the portion of its solutions that are multi-tenant cloud enabled. As the vendor expands its cloud capabilities and realizes cross-selling opportunities between its solutions, the overall value it delivers to customers should continue to improve.

Deltek

Deltek is a Leader in the 2018 edition of the ERP Value Matrix, focusing squarely on projectcentric businesses. The vendor delivers vertical solutions in several industries including architecture, engineering, and construction (AEC), marketing agencies, government contractors, management consulting, law firms, aerospace and defense, and non-profits. Since the last Value Matrix, Deltek has made three strategic acquisitions to extend its coverage in select verticals. First, Deltek acquired WorkBook to bolster its creative and marketing agency offerings (Nucleus Research, S98 — *WorkBook offers agency management visibility and insight*, June 2018). Second, to complement its GovWin product, Deltek purchased Onvia, which aggregates state, local, and education contracting opportunities. Lastly, Deltek acquired ConceptShare, which is a solution that caters to creative agencies by integrating online proofing with workflow automation.

In addition to the acquisitions, Deltek's development team has continued to produce a stream of improvements to its existing solutions. The vendor recently announced that its professional services solution will be renamed Vantagepoint upon its next release later this year. The solution is an end-to-end, project-centric ERP with vertical capabilities designed to support consulting and AEC industries throughout the entire project lifecycle. Deltek plans to continue increasing the global reach of the product with more localizations as well as integrating product information management capabilities.

On the innovation front, Deltek has invested in several technologies that customers are using today, including mobile functionality with Deltek Touch, integrated social capabilities that reduce the need for email, and data and key performance indicator (KPI) visualizations and dashboards that provide visibility across organizational silos.

With over 60 percent of its customers in the cloud, Deltek has also invested heavily in ensuring a secure environment to meet the needs of government contractors that handle sensitive or classified information. Additionally, the vendor has begun investing in capabilities that will further ease of use and increase automation such as AI and ML, natural language processes, and wearable technologies. As the vendor consolidates its new acquisitions and continues to deliver increased visibility and control to project-centric businesses, its ability to demonstrate value and positioning in the Value Matrix should improve.

Epicor

Epicor is an Expert in the 2018 edition of the ERP Value Matrix. The vendor delivers deep industry-specific functionality to a myriad of sub-verticals in retail, distribution, manufacturing, service industries, automotive, and lumber. Offering customers a full range of deployment options for its products (public or private cloud, hybrid cloud, and on premises), Epicor is expanding its partnership with Microsoft to deploy Epicor ERP products on Microsoft Azure cloud platform. As part of the vendor's strategy to accelerate cloud adoption amongst its customer base, Epicor is looking to leverage Microsoft's cloud technologies such as IoT, AI, and ML.

At its user conference earlier this year, Epicor outlined its vision for delivering Industry 4.0 capabilities, previewing, among other enhancements, a new user experience (UX) called Epicor Kinetic Design, which should start to be available with the vendor's next software release (Nucleus Research, S87—*Epicor focuses on users at Insights 2018*, May 2018). Epicor also announced a new independent software vendor (ISV) program designed to make it easier for partners to develop, market, and support extensions, as well as leverage the new technologies Epicor is bringing to its platform. To further aid in its products' extensibility, Epicor is partnering an application programming interface (API) integration platform vendor, Jitterbit, making the integration of cloud or on-premises applications simpler.

Having articulated its vision and charted its plans to modernize its solution offerings, Epicor must now execute on those intentions to keep pace with the market and deliver value to its customers. We expect Epicor's new UX to be a significant value-driver as customer adoption begins and accelerates, thereby improving the vendor's positioning in future editions of the Value Matrix.

FinancialForce

FinancialForce is a Facilitator in the 2018 ERP Value Matrix, delivering a cloud-based ERP built on the Salesforce platform. With a customer-centric approach to the visibility FinancialForce provides its customers, the vendor focuses on financial management, professional services automation (PSA), and people management with its core product offerings. A significant driver of its value, FinancialForce is natively integrated with Salesforce, providing Salesforce users with a familiar look and feel as well as seamless workflows from opportunity to invoice.

Since the last Value Matrix, FinancialForce has continued to work with Salesforce to leverage the platform's Einstein Analytics technology. At its annual user conference, FinancialForce announced it is delivering embedded reporting and analytics with FinancialForce PSA Analytics, which is powered by Einstein Analytics and gives users real-time visibility of KPIs across different service groups. Additionally, after last year's announcement that FinancialForce was partnering with ADP to provide HCM capabilities to the platform, the vendors have deepened their collaboration. Customers can integrate ADP's employee records with FinancialForce PSA so updates and changes are automatically shared between the two systems, with ADP acting as the primary system of record for employee data.

Though its investments in automating professional services business process are sure to generate additional value for customers, FinancialForce's ERP offerings have remained narrowly focused while other vendors in the market have diversified their cloud offerings, adding vertical industry-specific solutions to better serve their customers. While the native integration with Salesforce facilitates a customer-centric view of the customer's business, developing capabilities that answer the needs of sub-verticals within the professional services market would help FinancialForce's functionality score in future editions of the ERP Value Matrix.

 ## ROI Case Study: FinancialForce

Hewlett Packard Enterprise (HPE) · ROI: 158%

HPE deployed FinancialForce PSA (Professional Services Automation) as part of an initiative to consolidate its services business technology. The company leveraged FinancialForce PSA software to increase employee productivity and automate tasks. HPE was also able to collect and analyze accurate project data, helping managers make better business decisions and improve customer service.

THE COMPANY

Hewlett Packard Enterprise (HPE) is a global information technology company. HPE's solutions cover infrastructure and cloud solutions, IT security solutions, datadriven insights, mobile and connected device solutions, and business productivity. HPE also offers IT transformation services that help businesses along their digital journey. HPE's product portfolio includes storage, servers, integrated systems, networking devices, and software solutions. HPE is headquartered in Palo Alto, California.

CUMULATIVE NET BENEFIT

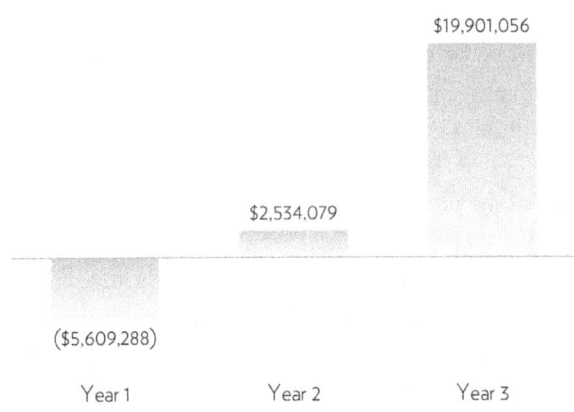

$19,901,056

$2,534,079

($5,609,288)

Year 1   Year 2   Year 3

KEY BENEFIT AREAS

HPE implemented FinancialForce PSA replacing several legacy systems to increase project visibility. First and foremost, HPE leveraged FinancialForce's capabilities to automate processes related to its project management workflows and increase worker productivity. Additionally, by consolidating over 15 different tools that were previously used to perform the service planning and delivery management functions, fewer IT resources were used and program stability was improved.

With FinancialForce PSA, HPE is now able to track project status from creation to completion. As a result, the Service Request Management division can be more proactive in its service delivery and gain insights on resource utilization and available capacity through better data collection and reporting. Finally, HPE is able to engage its customers more proactively and deliver better service due to the comprehensive project tracking and management capabilities made possible with FinancialForce PSA.

BEST PRACTICES

HPE developed dedicated teams to assist with change management as users adjusted to the new platform. The initiative helped limit the number of IT resources needed to make adjustments and recommend changes to improve the user experience. With FinancialForce PSA, HPE has the ability to deliver consistent and repeatable service execution and to differentiate services based on obligation type. HPE can better forecast service delivery and identify potential operational issues earlier. If issues occur, HPE can make the necessary adjustments due to better availability of data and the closed loop operations cycle.

By selecting FinancialForce, HPE took advantage of the consistent data model and native integration with the Salesforce platform creating a consistent user experience between the project management and CRM solutions. Users required less training time and could move seamlessly between applications.

· · · · ·

IFS

IFS is an Expert in the 2018 ERP Technology Value Matrix, delivering a new version of its ERP applications, IFS Applications 10, which includes a new UI called IFS Aurena. IFS also announced IFS Field Service Management 6, featuring a re-engineered user experience that is browser-based and has an improved scheduling engine. The vendor's capabilities are geared toward specific industries including automotive, aviation and defense, energy, high tech, engineering and infrastructure, oil and gas, industrial and process manufacturing, services, and retail.

With IFS Applications 10 and the Aurena interface, IFS features a bot that leverages AI to perform a suite of tasks using natural language processing. Users can access the bot from a host of messaging services such as Skype and Facebook Messenger. The latest edition of the software also includes capabilities to support service-centric organizations as well as demand-driven material requirements planning (DDMRP) that help companies reduce carried inventory and lead times. Additionally, IFS is looking to better support global customers with enhanced global tax management and multi-company capabilities.

IFS gives its customers choices in how they consume and deploy the software, offering cloud and on-premises models as well as subscription or perpetual license purchase options. Although IFS has a cloud-first strategy, many of its customers are opting for onpremises implementations. As the vendor works to consolidate the acquisitions it has made, particularly in field service management, customers should expect its solution suite to offer more cloud services and better value.

Infor CloudSuite

Infor continues as a Leader in the 2018 ERP Value Matrix, having made strides to integrate some of its recent acquisitions into a comprehensive cloud platform. Infor has maintained a vertical focus to its offerings, with its CloudSuite solutions covering a range of customer sizes and a plethora of industries, including but not limited to: aerospace and defense, automotive, equipment, fashion, distribution, food and beverage, healthcare, industrial machinery, manufacturing, and public sector. Designed to fit the customer's deployment needs, Infor offers its products as SaaS, hybrid deployment, or on premises, with the ability to scale up to serve global enterprises operating on multi- or single-tenant instances.

After acquiring a leading analytics and business intelligence (BI) vendor in Birst, Infor is inserting the capabilities into its technology stack as a BI platform in order to remove many of the data siloes that enterprise customers have (Nucleus Research, R76 — *Infor acquires Birst to access full value of data*, August 2016). On top of Birst and its data repository, Infor delivers a suite of technologies as part of Infor OS, such as Infor Ming.le, Infor Data Lake, Infor IoT, and Infor Coleman. With Coleman, the vendor is pushing forward AI with its digital assistant technology, which was announced at Inforum

2017. Combining third-party and standard Infor data connections like its commerce network, GT Nexus, Infor delivers an end-to-end solution suite with the flexibility for extensibility to cover the last-mile functionality that customers require.

IQMS

IQMS is a Core Provider in the 2018 ERP Value Matrix, focusing on mid-market manufacturing customers operating discrete and batch processes. With an end-to-end solution that covers ERP, MES, Quality Management Systems (QMS), and WMS, IQMS is underpinned by a single database and offers seamless, unified information flows for 21 types of manufacturing. IQMS provides customers a choice in how they deploy the software, including on premises, hosted perpetual licenses, and SaaS.

To ensure its customers get the most value from their shop floor, IQMS is deploying a suite of technologies that assist with production and process monitoring. Characterized as "Practical Industry 4.0", IQMS is looking to let companies improve their shop floor operations with capabilities such as backflushing operational information, overall equipment effectiveness (OEE), maintenance, repair and operations (MRO) tracking, and automation. IQMS is bringing several enhancements to its software including an integration layer that has standard interfaces with many major software vendors such as SalesForce, SAP, Oracle, and Marketo. IQMS is also delivering track and trace capabilities to support food and beverage, medical, and automotive customers.

With its MES solution, IQMS is targeting divisions of Fortune 1000 companies that frequently run SAP or Oracle for ERP but need a production and process monitoring solution, specializing in servicing plastic processing, metal forming, packaging, and assembly customers. As IQMS continues to offer functional depth and has begun to deliver Industry 4.0 technologies, additional usability enhancements should help the vendor keep pace with the market overall and maintain its position in future editions of the Value Matrix.

 # ROI Case Study: IQMS

KKSP Precision Machining · ROI: 93%

KKSP Precision Machining deployed IQMS EnterpriseIQ to improve shop floor control and production scheduling capability. With multiple production facilities and complex inventory routing, KKSP has improved data integrity throughout the manufacturing process and organization as a whole, since the system has been integrated. Nucleus found that IQMS enabled KKSP to improve the quality it delivers to customers, realize significant reductions in inventory levels, and improve bottom-line profitability through cost of quality reductions.

CUMULATIVE NET BENEFIT

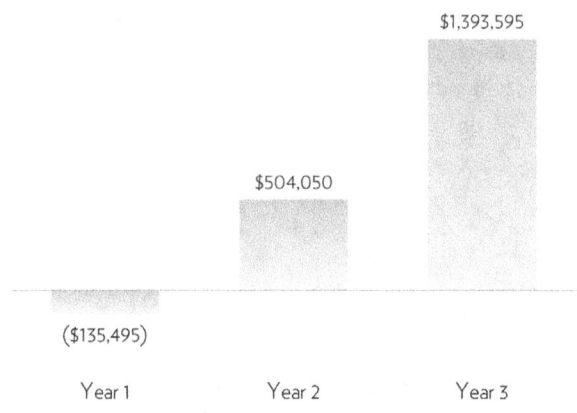

$1,393,595

$504,050

($135,495)

Year 1   Year 2   Year 3

THE COMPANY
KKSP Precision Machining produces screw machine parts. With four locations and hundreds of screw machines, KKSP is one of the largest manufacturers of its kind in North America. Founded in 1968, the company has a long history of producing parts for customers in industries such as automotive, industrial equipment, consumer goods and appliances, and medical devices. Producing over one million parts per day, KKSP can match its customers' specific

requirements while meeting all international standards for manufacturing and quality.

KEY BENEFIT AREAS

After deploying IQMS, KKSP was able to streamline its production schedule, reducing inefficiencies and safety stock. With more accurate BOMs, KKSP no longer needed as much safety stock as previously required. KKSP technicians could better predict what materials they needed to complete orders because the accuracy of the work schedule was improved. KKSP employees experience less idle time due to the accuracy of the production schedule. Rather than correcting bad data and chasing the wrong parts and work center requirements, employees can prioritize and plan jobs with greater accuracy.

With visibility of in-process jobs provided by IQMS, KKSP has a better understanding of the shop floor status, helping supervisors better manage operations. Finally, using the production schedule enabled by IQMS, KKSP is able to win business from larger customers. The greater accuracy of the BOM means KKSP can cost and price its products more competitively, winning business deals while still maintaining profitability and avoiding business with a higher risk profile.

BEST PRACTICES

KKSP had its information technology employees develop a system to monitor data quality that comes into IQMS on a continual basis. With a limited staff devoted to managing its technology ecosystem, KKSP needed a way to ensure users could quickly and easily understand if there was a data quality issue in the BOM or inventory source file for example.

KKSP's implementation process of IQMS ensured that the system could meet the company's needs. By starting with the smallest of its facilities and rolling it out to the rest of the organization, KKSP made sure any issues were resolved before expanding the deployment. For example,

KKSP initially had issues with scheduling which it addressed with the help of a local manufacturing implementation specialist.

. . . . .

Microsoft Dynamics 365 Business Central

Microsoft is a Facilitator with its Dynamics 365 Business Central product in the 2018 edition of the ERP Value Matrix. Its cloud-first, SMB offering, Microsoft Business Central is built on the NAV platform and was formerly called Dynamics 365 for Finance and Operations, Business Edition. With easy integration to Microsoft Office, Business Central offers a comprehensive suite of capabilities geared toward smaller companies such as financial management, SCM, project management, operations management, sales and customer service, and compliance.

With its October 2018 release, Business Central will boast a number of enhancements that stand to benefit both customers and partners (Nucleus Research, S122—*Microsoft Business Central Update*, August 2018). Features planned for the update include a new desktop UI, customizable Power BI reports within the Business Central homepage, and flexible deployment options. Microsoft is also extending connectivity to the intelligent edge via Power BI, PowerApps, Flow, and AI to its on-premises Business Central customers as well as Dynamics GP, Dynamics SL, legacy Dynamics NAV users. The move to bring legacy SMB customers to adopt intelligent cloud technologies will also allow partners to deliver migration services and should better prepare customers to make the leap to cloud when the business case presents itself.

Microsoft Dynamics 365 for Finance and Operations

Microsoft is a Leader in the 2018 ERP Value Matrix with its enterprise solution: Dynamics 365 for Finance and Operations. Continuing with its vision of unifying people, products and customers through digital feedback loops that are enabled by the technologies it is deploying, Microsoft is helping organizations along their digital transformation path. Dynamics 365 for Finance and Operations delivers capabilities around several pillars such as manufacturing, logistics and fulfillment, financial management, retail and commerce, planning, and product design.

In its Spring 2018 release, Microsoft provided improvements in analytics, data integration, extensibility and customization, supply chain optimization, and compliance. Microsoft's value proposition stems in part from the integrations it provides with other products in its ecosystem, including field service, project service automation, retail, sales, talent, and customer service. Underpinning and extending Microsoft's applications are capabilities such as Power BI, PowerApps, Common Data Service, and Microsoft Flow, which give users the tools to tailor their environment. With another release scheduled for October 2018, Microsoft gives its customers and partners a preview of what's to come and whether the capability will be delivered to cloud or on-premises deployments. Imbuing its applications with intelligent technologies such as AI has been a theme for Microsoft in recent releases, as the vendor seeks to improve efficiency and productivity by having the system make intelligent recommendations. As the vendor increases the number of use cases for these capabilities and their adoption becomes more widespread, its ability to deliver value that differentiates its products will improve.

Oracle ERP Cloud

Oracle is a Leader in the 2018 edition of the ERP Value Matrix, delivering an end-to-end enterprise cloud solution. With 140 modules in its suite, Oracle ERP Cloud addresses 23 industry verticals, including financial services, public sector, professional services, distribution, manufacturing, technology and communications, and higher education. Oracle also addresses all parts of the enterprise, covering financials, accounting, project financial management, revenue management, project management, procurement, analytics, supply chain, and risk management.

Delivering the entire cloud technology stack (i.e. infrastructure-as-a-service, platform-as-aservice, and software as a service), Oracle has been busy rewriting many of its applications to be cloud-native. In addition, the vendor supports the functional needs of global enterprises, offering SaaS solutions for human capital management (HCM), enterprise performance management (EPM), customer experience, and supply chain management (SCM) on an integrated platform. While the focus of Oracle continues to be on its cloud applications, customers running on-premises environments of Oracle Peoplesoft, E-Business Suite, and JD Edwards can expect the status quo to be maintained.

Oracle has also been investing heavily in Industry 4.0 technologies such as blockchain, IoT, chatbots, and AI (Nucleus Research, S38 — *Oracle Modern Finance showcases future of ERP*, February 2018). With its AI capabilities, one of Oracle's goals is to develop adaptive intelligent applications that are purpose-built and deliver value out-of-the-box. The intelligent applications, which have reinforced learning algorithms tuned to specific tasks, are designed to increase the amount of automation and agility customers have with capabilities such as workflow automation, continuous financial close processing, and automated compliance. Rather than a large product release around AI and ML, Oracle is embedding the technology in its applications, making customer adoption easier and speeding the time to value. Additionally, Oracle is developing paths to the cloud for its customers. For example, the Oracle Soar program announced in June 2018 is designed to automate many of the steps typically required when migrating to the cloud for PeopleSoft, Hyperion, and E-Business Suite customers. As the vendor continues to help its customers speed the time to value when adopting its cloud solutions, its overall usability should improve.

Oracle NetSuite

Oracle NetSuite is a Leader in the 2018 edition of the ERP Value Matrix. With a focus on pushing its products globally since its acquisition by Oracle in July 2016, NetSuite delivers a host of vertical solutions covering industries such as advertising and marketing agencies, consulting, financial services, healthcare, distribution, logistics, retail, nonprofit, manufacturing, food and beverage, and education. Since the last Value Matrix, NetSuite has built upon its SuiteSuccess engagement model, debuting 14 new editions of the program including for commerce, technology services, food and beverage manufacturing, advertising, media and publishing, consulting services, and manufacturing.

At its recent conference, SuiteWorld, NetSuite announced it had developed a new product that was designed to enable customers to deploy e-commerce capabilities within 30 days. The solution, called SuiteCommerce, gives customers the tools to launch and manage an ecommerce site without needing IT development resources. Another area of investment for NetSuite has been analytics and integrating business intelligence in its platform. Building on its existing BI tools, NetSuite is deploying capabilities that leverage AI and ML to deliver better insights to customers. As the vendor continues to make investments that facilitate customers moving to the cloud and shows that its innovations are delivering value, NetSuite's position in the Value Matrix will improve in the future.

Plex

Plex is a Core Provider in the 2018 ERP Value Matrix, focusing on a suite of industries including high tech, electronics, industrial manufacturing, automotive, aerospace, food and beverage, and metal forming. With Plex Manufacturing Cloud, the vendor covers business process management with accounting and financials, enterprise management, CRM, HCM, procurement, and supply chain planning. Plex also handles manufacturing operations with planning and scheduling, inventory management, quality management, production management, and engineering program management.

Since the last Value Matrix, Plex has made several moves to improve the value it is delivering to customers moving forward. For example, in May 2018, Plex announced the release of Plex Mobile, a native mobile application that is designed to be a seamless extension of Plex Manufacturing Cloud. With role-based functionality, Plex Mobile looks to address specific tasks within manufacturing operations such as inventory management, shipping and receiving, HCM, and production management. Additionally, in July 2018, Plex announced that it had acquired DATTUS, Inc., an industrial IoT (IIoT) vendor. DATTUS brings with it three core capabilities, namely IIoT connectivity, data management, and data analysis. The steps that Plex is making demonstrate that the vendor is looking to equip its customers with technologies that improve efficiency and bring modern capabilities to mid-market manufacturing customers, thereby delivering better value moving forward.

QAD

QAD is an Expert in the 2018 edition of the ERP Value Matrix having made significant strides in usability through its Channel Islands UX. QAD delivers ERP for manufacturing customers across six industry verticals: consumer products, automotive, food and beverage, industrial, high tech, and life sciences. Built for the cloud, the QAD Channel Islands UX is the culmination of a multi-year investment initiative to deliver a modern ERP on the QAD Enterprise Platform (Nucleus Research, S91 — *QAD highlights*

Channel Islands at Explore 2018, May 2018).

Part of QAD's investment has focused on bringing technologies to its platform that will help customers digitize their operations and deploy capabilities such as IoT, blockchain, 3D printing, and AI. The vendor has begun delivering advanced technologies as business services that customers can use as part of the core QAD Cloud ERP offering and/or as integrated tools or services via the platform. Since the last Value Matrix, QAD has made a host of functional improvements to its products, including a new version of QAD Production Orders and an updated supplier portal. As the vendor moves more of its customer-base onto its enterprise platform, increasing adoption of its UI and technology enhancements, its ability to demonstrates the value of its cloud products and positioning in the Value Matrix should improve.

ROI Case Study: QAD Cloud

Imperial Tobacco · ROI: 72%

Imperial Tobacco shifted its existing QAD production environment to the cloud rather than upgrading and refreshing its on-premises infrastructure, reducing its project costs by over seven million pounds. With its current environment reaching its end-of-life, Imperial Tobacco built a business case to upgrade to the latest version of

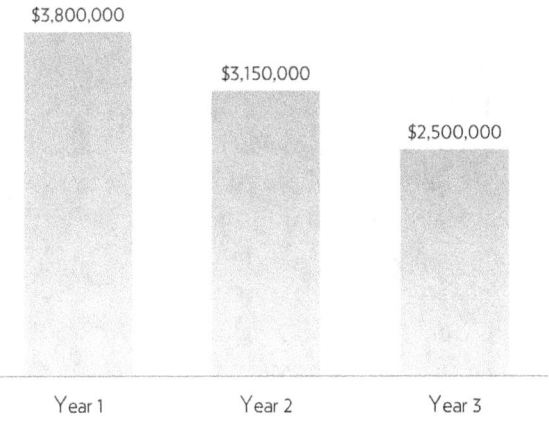

CUMULATIVE NET BENEFIT

$3,800,000	$3,150,000	$2,500,000
Year 1	Year 2	Year 3

QAD deployed on premises. However, by opting to lift and shift to the cloud, the company was able to extend the life of its current solution and lay the foundation for any future upgrade to the cloud. Imperial Tobacco was able to reduce its costs related to maintenance, development and technical staff, and technology hosting.

THE COMPANY

Imperial Tobacco is comprised of several manufacturing subsidiaries that produce and market a range of tobacco products and accessories. With a long heritage in the tobacco industry and locations in the UK, Poland, Russia, Germany, Spain, and Australia, Imperial Tobacco makes brands that are available in markets around the world including West, JPS, Golden Virginia, Davidoff, and Gauloises Blondes. Imperial Tobacco is one of five subsidiaries of Imperial Brands, which has diversified into next generation products as well as owning a traditional tobacco portfolio.

KEY BENEFIT AREAS

Imperial Tobacco realized benefits by shifting its QAD manufacturing environment to the cloud in a variety of areas, the first of which was reduced run costs. The business case on which Imperial Tobacco had started to execute to migrate to the latest QAD manufacturing environment on premises incurred costs of nine million pounds.

With the lift to the cloud of its current system, Imperial Tobacco reduced its one-off costs by seven million pounds. Furthermore, Imperial Tobacco reduced several of its operating costs including maintenance license costs and hosting fees. Imperial Tobacco was also able to reduce headcount among its technical and development staff by shifting management of its infrastructure to QAD's datacenter.

Finally, Imperial Tobacco significantly reduced its operational risk. Before relying on QAD to operate and manage its infrastructure, Imperial Tobacco had its datacenter at one of its production locations. By moving its manufacturing environment to a professionally run datacenter, Imperial Tobacco reduced the risk of damage and security of its infrastructure assets.

BEST PRACTICES

By lifting and shifting to the cloud, Imperial Tobacco was able to extend the life of its system for several years. The system had been nearing its end-of-life, with maintenance support ending. However, by moving to the cloud, the company guaranteed support from QAD for eight additional years. Additionally, by moving to cloud infrastructure, the company is now comfortable with continuing to use cloud when it deploys its next manufacturing solution.

· · · · ·

Qualiac

Qualiac is a Facilitator in the 2018 ERP Value Matrix. In October 2017, Qualiac was acquired by Cegid, a retail and talent management software vendor headquartered in France. The move is designed to allow Cegid customers to extend their solution ecosystem beyond payroll, taxation, and HCM. Catering to the SMB market, Qualiac has been operating as a subsidiary of Cegid; however, there are plans for the Qualiac brand to be consolidated under the Cegid brand at some stage.

Qualiac offers ERP solutions focused on finance, production, supply chain, and public sector institutions. It also provides complementary modules to aid in decision-making and collaboration, such as data-connectors with third-party BI tools and a cross-functional workflow information manager. The solution is highly configurable, requiring minimal IT resources for users to leverage the embedded analytics and data traceability that has helped Qualiac make inroads in the healthcare market. With Qualiac's acquisition and consolidation under the Cegid brand, the future of the solution and its activity in the market remains to be seen.

Rootstock

Rootstock is an Expert in the 2018 Edition of the ERP Value Matrix. Focused on manufacturing and distribution delivered in the cloud on the Salesforce platform, Rootstock services customers in high tech, machinery, electronics, industrial equipment, engineer to order, project manufacturing, and wholesale distribution. Since the last Value Matrix, Rootstock's biggest move was the acquisition of Kenandy in January 2018, which is an ERP cloud vendor built on the Salesforce platform. It is focused on manufacturing, supply chain, distribution, financials, and order-to-cash. The move consolidates Rootstock's leading position in the manufacturing and distribution space on the Salesforce platform. Rootstock has been investing in actualizing the technologies delivered on the Salesforce platform for its customers, such as Einstein Analytics and IoT. For existing Salesforce customers, the native integration and familiar Lightning UX are the value-drivers that justify consolidating significant portions of the enterprise technology on the Salesforce platform. Rootstock has been successful in attracting customers who aren't already Salesforce customers, demonstrating the value of its technology that is not underpinned by Salesforce. As the vendor continues to expand beyond the Salesforce umbrella, its ability to demonstrate its stand-alone value and its position in the Value Matrix should improve.

Sage

Sage is a Core Provider in the 2018 edition of the ERP Value Matrix. Delivering solutions for small- and medium-sized businesses as well as enterprises, Sage covers industry-specific functionality in distribution for wholesale, logistics, and retail; manufacturing for high-tech, industrial equipment, metal fabrication, and medical devices; process manufacturing for chemicals, cosmetics, food and beverage, and pharmaceuticals; and services for advertising, consulting, engineering, IT/software, and equipment maintenance and repair. The vendor underwent a significant rebranding in October 2017, introducing Sage Business Cloud as the one-stop-shop for accounting, financials, people, and enterprise management.

As part of its shift to cloud, Sage is looking at how it can apply ML and AI to its product offerings in order to reduce administrative tasks. The vendor is also building up a comprehensive marketplace for its VAR and ISV partners, allowing customers to find the functional extensions that would benefit their business. The recent abrupt departure of Sage's CEO, Stephen Kelly, raises questions about the future trajectory of the company and where its investments will deliver value to customers moving forward. With a large existing install-base, Sage needs to provide a coherent path to the cloud and modern ERP solutions or its risks further loss of legacy customers to rival vendors.

SAP Business ByDesign

SAP Business ByDesign is a Facilitator in the 2018 edition of the ERP Value Matrix, serving SMBs with its born-in-the-cloud solution. Designed as an end-to-end ERP, Business ByDesign supports finance, project management, procurement, supply chain management (SCM), customer relationship management (CRM), and human resources. The solution delivers industry-specific functionality to organizations in professional services, wholesale distribution, manufacturing, and the public sector (for North American customers only). With a global presence, Business ByDesign offers language, currency, and regulatory localizations for over 130 countries.

In the most recent release of the product, SAP Business ByDesign 1808, SAP took steps to bring more intelligent capabilities to the platform. Users can integrate predictive analytics with information delivered to their tailored overview pages, drawing on pre-packaged content from SAP. The solution leverages SAP Analytics Cloud to gain insights into the drivers of the customer's business. SAP is focused on keeping the user experience simple while also providing flexibility, so customers can tailor the solution to their exact needs (Nucleus Research, S8—*SAP Business ByDesign deliver extensibility*, January 2018). The vendor plans to deliver more out-of-the-box capabilities with each release to make user adoption easier and shorten the time to value realization for customers. We expect further functional enhancements to improve SAP Business ByDesign's position in future editions of the Value Matrix.

SAP S/4HANA

SAP is a Leader in the 2018 ERP Value Matrix, having made strides to bring more of the capabilities that deliver intelligent ERP to SAP S/4HANA and S/4HANA Cloud. Continuing its cloud-first strategy, SAP is releasing quarterly updates to its S/4HANA Cloud products, while the on-premise version receives yearly major releases. With each update of its solution,

SAP combines and prepackages more of the technologies enabled by SAP Leonardo, such as machine learning, conversational AI, and predictive analytics, with the 25 industry verticals SAP's ERP services.

In the latest release, the vendor already provides 33 scenarios out-of-the-box in total. Some noteworthy scenarios SAP added most recently include: ML-informed goods receipt/invoice receipt account reconciliation, where the system learns how to handle new business scenarios based on past decisions and provides recommendations to the user to significantly speed up and increase the accuracy of the monthly book closing process; project forecasting with ML analysis of historical project data as part of enterprise portfolio and project management to make predictions based on facts rather than individual judgments; and demand-driven replenishment with ML analysis of buffer levels and adjustments designed to improve inventory management by operating based on real demand and reducing safety stocks.

By bringing intelligence to its applications, SAP is looking to reduce or eliminate mundane, repetitive tasks. Instead of a technology looking for a problem, SAP is targeting use-cases where customers can realize value quickly. As a result, scenarios that leverage machine learning can be role-specific, which makes it simple for the user to take advantage of the technology. SAP is looking to demonstrate sufficient benefits to entice more customers to make the move to cloud, which will enable customers to more quickly consume the latest capabilities the SAP provides in its updates. As SAP expands the number of scenarios in which it is applying intelligence, it will continue to increase customer efficiency and the value S/4HANA Cloud delivers.

 ## ROI Case Study: SAP S/4HANA

Centria · ROI: 29%

Centria deployed SAP S/4HANA to unify its enterprise resource planning (ERP) solution landscape as well as simplify business processes and implement mobile solutions. By consolidating three instances of SAP, Centria was able to improve the speed at which it performed its financial close and reporting. With SAP S/4HANA, Centria avoided software development costs and reduced its infrastructure spending. By leveraging the SAP Fiori user experience, Centria is positioned to deploy multiple mobile applications for its users, which will further improve the efficiency of its operations.

CUMULATIVE NET BENEFIT

($76,666)

($220,366)

($313,598)

Year 1   Year 2   Year 3

THE COMPANY
Centria Servicios Administrativos S.A. is a professional services company focused on information technology, finance, logistics, business risk, legal, and comptroller services. Based in Lima, Peru, Centria is part of a conglomerate of companies operating in a wide range of industries, including banking, fishing, mining, cement manufacturing, insurance, and healthcare. Offering both consultative and managed services, Centria works primarily with other companies within the Breca Group.

KEY BENEFIT AREAS
Deploying S/4HANA enabled Centria to unify its fragmented ERP solution, simplify accounting processes, and implement mobile solutions. Initially, Centria had considered partnering with another software company to connect web applications to its legacy version of SAP ERP. The third-party developer had scoped Centria's needs and delivered project proposals that Centria was considering before deciding to upgrade to S/4HANA. Centria realized

that it could accomplish its mobile development goals with S/4HANA at a reduced cost.

Additionally, by leveraging the S/4HANA database, Centria reduced its spending on data compression. The company had previously made use of both IBM and SQL's databases, which it was able to eliminate with the storage efficiency gained through S/4HANA.

Moving to S/4HANA also simplified accounting processes and enabled Centria to reorganize its staff, reducing the number of accountants it required to perform its month-end close and prepare its financial reports.

Finally, management is more productive and can spend less time on infrastructure administration. Accountants are also more productive due to simplified accounting processes such as invoicing, and reduced report creation time.

BEST PRACTICES

Initially, Centria was looking to build four or five solutions that connected to its SAP system. Instead of using a third-party to build the web applications, the company determined it could achieve the same results at a lower cost. More importantly, Centria will be able to expand its use of the S/4HANA system and Fiori user experience to additional applications, as it has only 15 percent of its functionality in Fiori currently. The company also plans to expand its use of S/4HANA and Fiori to other parts of the organization, further consolidating its enterprise landscape.

With plans to develop 15 to 25 applications for its users each year, Centria has a platform on which it can deliver more modern capabilities and systematically transform its enterprise technology as its needs change. In addition to Centria benefitting from the new application, it can extend its business model to deliver new value-add services to its clients.

· · · · ·

SYSPRO

SYSPRO remains a Leader in the 2018 ERP Value Matrix after investing in a new User Interface that will be generally available with SYSPRO's July 2018 release. With a strong focus on manufacturing and distribution, SYSPRO serves verticals such as automotive parts and accessories, electronics, food and beverage, industrial machinery and equipment, fabricated metals, packaging, and plastics and rubber. SYSPRO has over 15,000 customers and provides localization support across 62 countries. The vendor has moved steadily upmarket and is able to service tier-one enterprises as well as small- and medium-sized businesses.

As part of its latest release, SYSPRO focused on delivering flexibility to its customers. With the release of SYSPRO's new web-based user interface (UI), called Avanti, customers can access the software from any web-browser as well as leverage SYSPRO's customizable mobile application, Espresso. Additionally, customers have the choice to deploy onpremises or to a private cloud as well as the option of purchasing software through subscription or perpetual licenses.

SYSPRO is bringing extended services to its software with capabilities that enable customers to leverage Microsoft Azure IoT hub and Microsoft Azure ML/AI bot technology. SYSPRO's bot, which can be deployed to any messaging platform, comes with 60 skills available to users out of the box, such as adding calendar appointments and reading emails. The vendor is also introducing a new way to consume ERP with its Harmony interface, which delivers ERP information in a similar manner to that of a social media feed, providing data and key metrics to users.

SYSPRO continues to demonstrate its commitment to delivering value to its customers, focusing on practical applications of the latest enterprise technologies. As it moves upmarket, SYSPRO has ensured that its product is able to effectively scale, making its value proposition competitive in the verticals it serves.

Unit4

Unit4 is a Facilitator in the 2018 edition of the ERP Value Matrix, scoring highly in usability, reflecting the vendor's focus on people and flexibility. The vendor delivers a host of solutions focusing on professional services, higher education, public services, and nonprofits. Its flagship ERP solution, Unit4

Business World, offers financial management, project management, procurement management, human resources and payroll, as well as field service and asset management as its core capabilities.

As part of its platform strategy, Unit4 is looking to enable customers to take advantage of new technologies facilitated by the cloud. For example, the vendor recently released an extensions kit that allows customers and partners to build low-code/no-code applications that can be shared on Unit4's cloud application marketplace. The capabilities are designed to allow customers to consume cloud-based services on top of Business World. Unit4 has also been investing in its digital assistant, Wanda, which works through any messaging application like Skype or Slack. The technology leverages natural language processing, pattern recognition to automate repetitive tasks, and links into a company's business systems to operate as the interface for the user, which replaces manual searches through disparate information repositories.

Offering SaaS, dedicated cloud, and on-premises deployment, Unit4 allows extensibility that increases its ease of use while providing functional depth in the verticals it serves. The vendor is investing in the concept of a self-driving ERP, which will automate traditional business processes and further the value that the solution delivers. We expects that Unit4's investments will increase the value customers realize through its usability and further set it apart from other vendors in the market.

VAI

VAI (Vormittag Associates, Inc.) is a Core Provider in the 2018 ERP Value Matrix, focusing on midmarket customers in several industries. The industries VAI services include electronics, plastics, medical products, automotive, industrial equipment, apparel, food and beverage, building and electrical supply, and plastics. VAI also has two industry-specific solutions covering pharmaceutical and food distribution and manufacturing. The vendor has a suite of cross-industry capabilities such as CRM, WMS, analytics, and mobile, in addition to financial management, distribution management, manufacturing management, and retail as part of the core product capabilities within S2K Enterprise.

Since the last Value Matrix, VAI has made a number of announcements centered on bringing modern technologies to VAI customers. For example, in October 2017, VAI unveiled its Enterprise Intelligence approach which leverages cloud, mobile, and analytics, and is designed to empower midmarket companies to make better business decisions, giving the right information to the right user at the right time. More recently, VAI announced a partnership with SAP Concur, a spend management solution provider. The partnership provides VAI customers with an integration between Concur's cloud application and VAI's enterprise platform. Designed to eliminate manual, time-consuming tasks associated with expense management, VAI customers can leverage the capabilities to gain real-time access to expense data as well as leverage the mobile application to input data from anywhere.

Offering customers unlimited user licenses, VAI can deliver value to fast-growing midmarket companies, especially when customers are going through mergers and acquisitions. As VAI shifts more of its customers to cloud and delivers more Industry 4.0 capabilities to its platform, the vendor's ability to answer the needs of its customers and its position in future editions of the Value Matrix should improve.

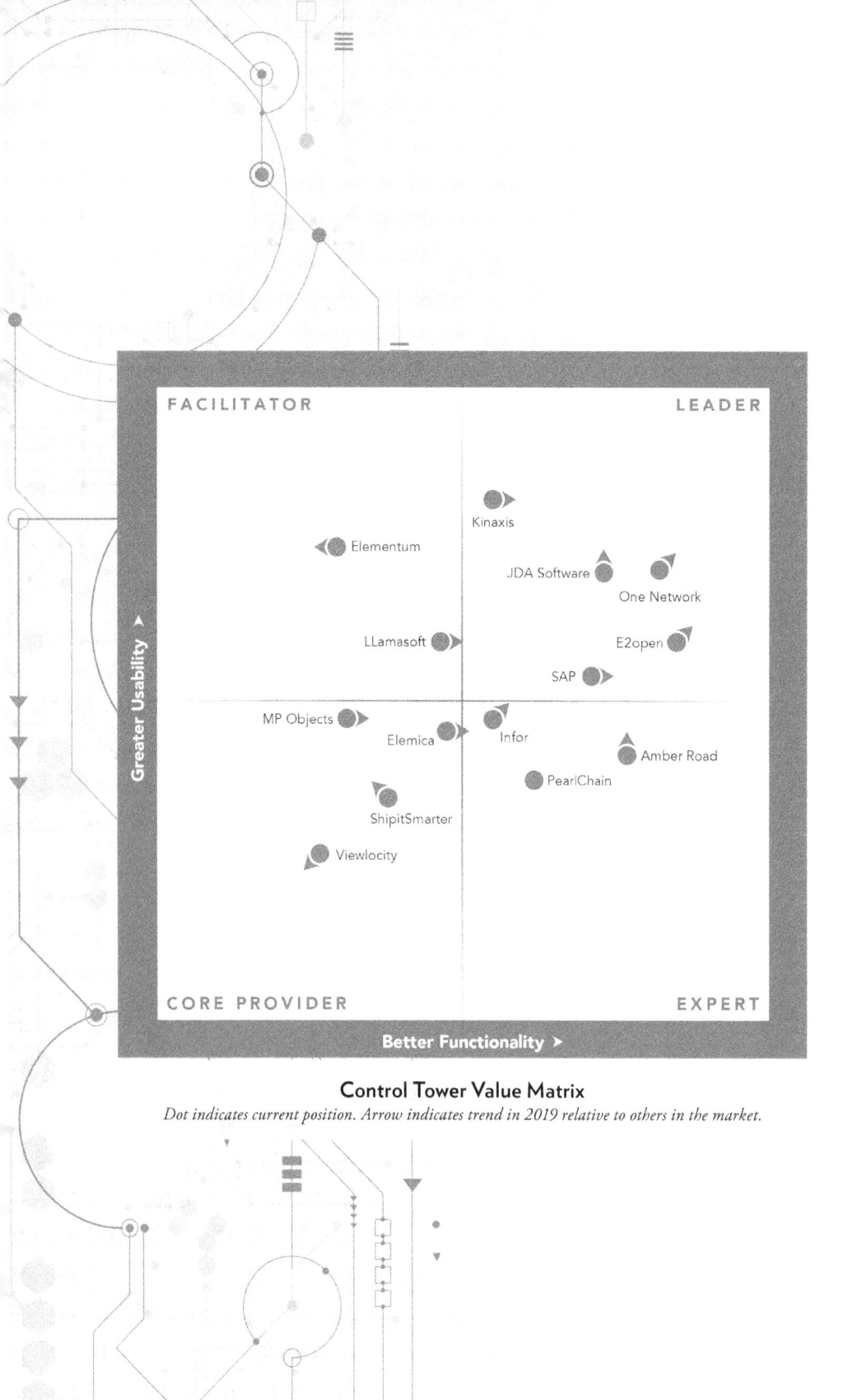

Control Tower Value Matrix

Dot indicates current position. Arrow indicates trend in 2019 relative to others in the market.

CHAPTER 7
CONTROL TOWER

Supply chain control towers are an increasingly vital tool for enterprises to gain visibility over their extended value chains. Striving to be the connective tissue between planning and execution, control towers are increasingly an avenue by which companies start to modernize and digitize their supply chains, cutting across organizational silos. By the nature of the solutions and their reliance on real-time data, control towers inherently require organizations to move away from spreadsheet planning and disparate, disconnected systems.

In terms of value delivery, the bar for control towers is relatively low, since the solutions' visibility is unprecedented in most organizations, where business processes don't cut across traditional operational silos. Organizations can achieve better collaboration with trading partners and logistics providers as well as identify potential issues sooner, allowing supply chain planning and execution to function more proactively rather than reactively.

In this Value Matrix, we evaluated the control market vendors based on their product usability and functionality and the value that customers realize from the capabilities of the product. As a snapshot of the control tower solution market, this research is intended to help inform consumers about how well vendors are delivering value to customers, and what a prospect can expect moving forward based on the investment vendors are making today.

In 2018, some supply chain control tower vendors delivered tools that surpassed the ability of customers to adopt the technology. Along the overall supply chain maturity curve, we found that the majority of customers are still relatively low in their readiness to adopt technologies like machine learning and artificial intelligence (AI), which vendors are looking to bring to their solutions. With an understanding that customers are on the "crawl, walk, run" continuum, vendors have sought to build tools that reduce the barriers to acceptance and address low-hanging fruit.

We found that tools that help adoption and ease change management for users are drivers of significant value. Change within the supply chain has proven incredibly difficult for many organizations, with the metrics of success often based only incremental improvements rather than wholesale transformations. Leading vendors in the 2018 edition of the Control Tower Technology Value Matrix are making it easier for companies to use their technology by delivering the right information to the right user at the right time.

· · · · ·

Amber Road

Amber Road is an Expert in the 2018 edition of the Control Tower Value Matrix, specializing in moving goods across borders. Amber Road provides visibility across a customer's execution network. Operating as a one-stop-shop for global trade management, Amber Road enables customers to build a digital version of their supply chains, thereby connecting customers with overseas suppliers, logistics providers, freight forwarders, customs brokers, and carriers. The software is supported by dynamically changing content around trade rules and tariffs which are gathered, translated, and interpreted by in-house specialists. Both internal stakeholders and external partners are linked through a common platform allowing users to collaborate and implement controls.

In addition to execution visibility, customers can leverage the software for global sourcing, risk and quality management, transportation management, production management, and import and export management. The solution delivers many capabilities out-of-the-box, but for customers that need additional features, Amber Road provides configurability that allows for customizations. An API integration layer enables customers to connect external, thirdparty, and IoT data they need to model their operations. Amber Road has invested in automating the collection of data and reducing manual tasks, so users can manage by exception on a global basis. The vendor is investing in delivering global inventory visibility capabilities, total landed cost management, and origin management. To improve usability, the vendor is working on improving process orchestration and workflow as well as reporting capabilities. Setting the standard for execution visibility, Amber Road's current investments should improve its position in future editions of the Value Matrix.

E2open

E2open continues as a Leader in the 2018 Control Tower Value Matrix, having made a series of acquisitions to extend its capabilities in downstream execution. Geared toward global enterprise customers with extended value chains, E2open brings together its different planning and execution solutions with a consistent user interface. It is also underpinned by a multi-enterprise network that delivers a single consistent data model.

Over the last few years, E2open has made a series of acquisitions designed to extend its capabilities beyond its beginnings in planning. Of late, downstream logistics has been the focus, with E2open purchasing the transportation management vendor, Cloud Logistics, and ocean shipping vendor, INTTRA (Nucleus Research, s166—*E2open expands downstream*, October 2018). Consolidated under its Harmony user interface, E2open is looking to offer visibility from end-to-end as well across trading partners outside the enterprise. E2open has been investing in capabilities that will allow customers to perform root cause analysis on errors. The vendor is also investing in delivering an activity stream that will integrate any unstructured activity that connects to a business process.

The vendor seeks to enable the convergence of supply chain management between the external ecosystems that surround an enterprise with the planning and execution systems that E2open provides. With the vendor still in acquisition mode, customers should expect E2open to continue to fill out their functional needs.

Elementum

Elementum is a Facilitator in the 2018 Control Tower Value Matrix. The vendor's control tower is called Orchestration Platform and provides multi-enterprise visibility. The solution is delivered as software as a service and tracks actions, participants, and decisions across workflows. Designed to layer over transactional systems, the solution is highly configurable and delivers flexibility that helps customers achieve value.

The vendor has been investing in artificial intelligence technologies designed to assist decision making. Capabilities include natural language processing and predictive analysis for shipments, inventory, and supply chain errors. The solution also allows automation of routine errors to reduce manual interventions. Incident Management platform learns the workflows in order to successfully resolve future issues. Working in tandem with Elementum Product Graph, customers can digitally map their global supply chains' connecting nodes and track shipment in real-time. Continued investments in functionality should help the vendor's position in future editions of the Value Matrix.

Elemica

Elemica is a Core Provider in the 2018 edition of the Control Tower Value Matrix. With the visibility part of its solution set, Elemica provides three applications: Elemica Pulse, Elemica Trace, and Elemica Risk. Elemica Pulse provides end-to-end visibility and monitoring. Combined with Elemica Digital Supply Network, Elemica Pulse establishes rules-based event management and machine learning analysis of exceptions. Pulse also provides early alerts to discrepancies and other errors.

Elemica Trace, which was released in March 2018, provides shipment visibility in real-time, collecting signals from carriers and shipment data to predict delivery and identify potential issues. The solution also allows the user to manage by exception, with email alerts and notices issued when a shipment is delayed.

Elemica Risk leverages a partnership with the carrier, DHL, by integrating Risk with DHL Resilience 360, helping users track incidents and connect them to the supply chain impact.

By connecting issues to the customers' granular data, such as purchase orders, users can understand the impact on individual stakeholders and take the appropriate action to rectify the situation. Having undergone a major rebranding overhaul and with its new product capabilities, Elemica is on a path to improve the value it delivers and its position in future editions of the Control Tower Value Matrix.

Infor

Infor is an Expert in the 2018 Control Tower Value Matrix. Since its acquisition of GT Nexus, Infor has been investing in expanding the solution beyond its foundation in global trade management. After going quiet as to the future of the solution, Infor announced the next generation of its control tower solution, called Control Center, which is scheduled to become generally available early in the coming year. Building on the capabilities GT Nexus delivered, Control Center will feature a single instance multi-enterprise business network and combine the analytical capabilities of Infor Coleman with supply chain execution.

Control Center will introduce a new user experience and bring the intelligence of Coleman machine learning to help users with predictions about what will happen based on analysis of historical and real-time data. Early adopter customers are currently tracking shipments with IoT capabilities to better predict delivery times, with truck tracking coming soon. With a long heritage of supporting global supply chains with enterprise-to-enterprise visibility, Infor's new solution will have improved usability, which should help customers achieve value and improve the vendor's positioning in future iterations of the Value Matrix.

JDA Software

JDA Software continues as a Leader in the 2018 Control Tower Value Matrix. The vendor delivers control tower capabilities with its supply chain planning applications. JDA's applications sit on its platform technology that allows development and integration of thirdparty services and extension applications, as well as data and analytics services from JDA's new Luminate product extensions. Announced at its annual user conference, JDA Luminate is a suite of core product enhancements that are cloud-native, platform-agnostic microservices and components that leverage many of the technologies delivered by JDA's cloud partners in Microsoft and Google (Nucleus Research, s82—*JDA expands platform play at FOCUS 2018*, May 2018).

JDA's strategic goal is to deliver an autonomous supply chain that relies on visibility as its foundation but adds predictive and prescriptive analytics as well as leveraging contextsensitive self-learning capabilities that help orchestrate the supply chain from end-to-end. Until now, JDA's solution for control tower, the Agile Workbench, offered master planning visibility.

With its Luminate Control Tower, JDA is extending visibility into execution, thus enabling collaboration between internal and external stakeholders and execution workflow orchestration. The vendor has been proactive in establishing partnerships with third-party vendors that provide relevant data to its control tower, such as real-time weather, transportation, and track-and-trace. As more customers move to the Luminate Control Tower, we expect JDA to further push the boundaries on the visibility and control that customers can realize.

Kinaxis

Kinaxis is a Leader in the 2018 version of the Control Tower Value Matrix with its RapidResponse platform focused on real-time planning. Providing robust scenario and what-if analysis, Kinaxis has been building out industry-specific functionality focused on high tech, life sciences, automotive, industrial, consumer packaged goods, and aerospace and defense. Kinaxis has been investing in the connectivity of its platform to get data in and out of the planning system more easily and provide better visibility across the organization to users.

RapidResponse delivers value through its flexibility, which allows users to configure the user interface based on role and which key performance indicators (KPIs) are most relevant. Looking ahead, the vendor is investing in practical applications of artificial intelligence and machine learning technology that enable the Self-Healing Supply Chain. Kinaxis is focusing on ensuring user adoption of its innovations, starting with lead times and adding capabilities as users can accept them. An additional area of focus is further easing the integration process for RapidResponse. Kinaxis is looking to enable integration without predefined schemas and inferred relationship mapping.

The flexibility and ease of use of RapidResponse continues to be a major differentiator, with customers able to go live quickly and start realizing value. Supporting some of the largest global companies with complex value chains, Kinaxis is making investments today that will further cement its place in future editions of the Control Tower Value Matrix.

LLamasoft

LLamasoft is a Facilitator in this year's Control Tower Value Matrix.

Focused squarely on supply chain design, the vendor is bringing to market its Digital Design & Decision Center, which contains three core capability areas: Visualizer, Modeler, and App Builder. LLamasoft Visualizer enables users to analyze the performance of their end-to-end supply chain and make adjustments that optimize performance as well as track key metrics and identify potential issues. The solution's control tower capabilities give customers the ability to easily evaluate different scenarios based on service, cost, risk, and any other relevant KPIs.

With the mantra of "Planning by Design," LLamasoft's Data Hub, which delivers a global data model that enables users to operate on a single version of the truth, allows customers to map and visualize their value chain as well as perform extensive scenario analysis. Additionally, the Data Hub infrastructure enables collaboration and analysis across the entire value chain.

With a cloud-based architecture, LLamasoft customers frequently establish a Center of Excellence to get the most value out of the solution and its flexible use-cases, making use of the integrated platform to perform analyses of their supply chain configuration in order to optimize and recalibrate. Able to service some of the most complex supply chains for some of the world's largest companies, LLamasoft continues to expand its functionality and should expect to see its position improve in future editions of the Value Matrix.

One Network

One Network is a Leader in this year's edition of the Control Tower Value Matrix. The vendor delivers its control tower application on top of its multi-party business network, which has over 60,000 companies and 16,000 carriers on it. One Network has been investing in delivering autonomous intelligent agents as parts of its applications, which are designed to simplify and optimize transactions across the entire value chain. Able to offer decision support or execution automation, One Network's agents operate on subnets that partition the larger network, so the agents can optimize and replan in real-time. One Network delivers a single version of the truth for all network members, enabling buyers, suppliers, logistics, and planners to operate based on the same information.

One Network has specialized its solution to support automotive, retail, high tech, construction, pharma, and aerospace and defense, and provides applications that manage both planning and execution. Customers can plan and optimize across organizational silos with execution data fed back into the planning engine to ensure it is up to date. The vendor has been investing in bringing Blockchain technology — which supports Ethereum and Hyperledger — to its platform, providing private multi-party ledgers on which customers can develop modules and set permissions for trading partners. With many of its customers on the cutting edge of what control tower technology is delivering, One Network is helping transition customers from siloed legacy systems to a collaborative, real-time network.

PearlChain

PearlChain is an Expert in the 2018 Control Tower Value Matrix, delivering several areas of supply chain management to specific industries. The vendor specializes in managing the contractual obligations between parties and the complex fulfillment processes that are required. Focus areas include food and process manufacturing, construction, automotive, and ship management. The vendor is designed to connect planning and execution across sales and operations planning, manufacturing execution, warehouse management, contract management, shop floor control, purchasing, and order management.

The vendor's primary offering, called Core, offers visibility of shipments down to the SKU level, as well as dashboards based on roles. The solution takes real-time constraints into account when modeling the details of contractual agreements. Delivered as a cloud or onpremises offering, PearlChain Core provides a single version of the truth to users, with a service-based architecture that integrates with legacy ERP systems. The vendor has continued to deliver usability enhancements and additional services that improve the value that customers realize from the solution.

SAP

SAP is a Leader in the 2018 Control Tower Value Matrix. SAP's Supply Chain Control Tower integrates with its Integrated Business Planning (IBP) solution suite and combines real-time visibility and analytics that measure the impact of exceptions and help users make better decisions. Along with SAP IBP, the control tower solution can be deployed on SAP HANA Platform, as well as integrated with SAP Leonardo (for analytics) and SAP Ariba (for supply chain collaboration). To gain more complete visibility, the solution can draw on data from enterprise resource planning (ERP) systems, transportation management, and advance planning and optimizer (APO) as well as third-party and external systems.

The vendor has continued to invest in machine learning technology that is designed to generate decision recommendations for the user. In addition to decision-making support, the control tower solution provides collaboration capabilities with external trading partners as well as alerts based on exceptions to business rules. Based on the SAP Fiori user interface, customers are given dashboards that provide KPIs and real-time supply chain performance data. Designed to support the largest and most complex enterprises, SAP's Supply Chain Control Tower can help organizations mature their supply chain processes and improve overall performance.

ShipitSmarter

ShipitSmarter is a Core Provider in the 2018 Control Tower Value Matrix. Delivering cloudbased software as a service solutions, ShipitSmarter specializes in managing the entire shipment lifecycle from order creation to financial settlement. Along the shipment lifecycle, the vendor delivers capabilities addressing system integration, carrier management, transportation management, freight audit, and financial management. ShipitSmarter prides itself on its independence from the carriers with which it works, helping customers achieve value by functioning as a strategic sparring partner.

In August, the vendor announced a new business intelligence tool called Advanced Reporting. Designed to bring greater configurability and simplified dashboards to its toolset, ShipitSmarter's new service is responding to the needs of customers to be able to view accurate real-time data presented in digestible ways. As a best-of-breed vendor, ShipitSmarter delivers value to customers that are looking address their complex shipment management processes.

Viewlocity

Viewlocity Technologies is a Core Provider in this year's edition of the Control Tower Value Matrix. The vendor has a long history of delivering solutions focused on supply chain visibility and control. Viewlocity has a suite of solutions covering supply chain design, inventory and asset management, supply chain management, logistics and transportation, planning and production, order lifecycle management, and supplier management. All of Viewlocity's products are offered as SaaS, managed services, or license deployments. The vendor focused on several industries with its solutions, including automotive, aerospace, manufacturing, retail, high tech, consumer goods, and logistics.

Founded in 1999, Viewlocity is owned and managed by Fog Software Group, which is an operating group of Vela Software Group, which is an operating group of Constellation Software Inc., a global software service aggregator with groups operating in 60 market verticals. Since its acquisition, investments in Viewlocity's solutions have been minimal, with the controlling interests continuing to push the vendor's Control Tower Platform solution. Although it is able to service global customers, without further investment in new capabilities, the vendor will continue to lose ground relative to the rest of the market.

FACILITATOR   LEADER

Greater Usability ▲

Logility

LLamasoft

ToolsGroup

One Network

Kinaxis

E2Open

Blue Ridge

JDA Software

Vanguard

GAINSystems

Oracle

Infor

SAP

Manhattan Associates

4R Systems

Smart Software

Slimstock

CORE PROVIDER   EXPERT

Better Functionality ▶

Inventory Optimization Value Matrix

Dot indicates current position. Arrow indicates trend in 2019 relative to others in the market.

CHAPTER 8
INVENTORY OPTIMIZATION

In the 2018 edition of the Inventory Optimization (IO) Value Matrix, we analyzed solutions in the market that are designed to minimize inventory across the value chain while maintaining or increasing service levels for customers. Based on their ability to deliver value through the usability and functionality of their solutions, some vendors featured in this year's Value Matrix are focused on providing capabilities to their customers that increase both usability and functionality by allowing software to manage inventory levels and reduce human input.

While this Value Matrix continues to differentiate between solutions that offer inventory management rather than inventory optimization, we found a persistent trend among customers: the synergies between modules adjacent to the inventory optimization tool. Rarely purchased as a stand-alone module, customers saw increased benefit when IO solutions were informed most immediately by demand planning and pushed the optimal inventory levels to a supply planning module.

Since the last IO Value Matrix, vendors have focused on several areas to improve the usability and functionality of their offerings, such as improved modeling and visualizations, machine learning algorithms, and multi-enterprise collaboration. On the automation front, users are now able to set the inventory and service policies they wish to achieve and allowing the software to determine the optimal path forward with no human intervention necessary.

Looking forward, the inventory optimization vendors are continuing to look for ways to connect their planning tools more closely with point-of-sale (POS) data and other demand signals to make their solutions more responsive. Vendors are tasked with leveraging machine intelligence to better predict demand while giving the customers agility in reacting to issues which arise. These capabilities surround the inventory optimization engine, where automated best-fit model selection is a table stake and a more holistic view of inventory in the supply chain network is gaining more interest from customers.

In terms of usability, vendors are focused on better reporting and modeling tools to give users a better picture of where they stand instead of just providing a spreadsheet of numbers. User adoption is also undergoing a shift, as vendors are investing in ways to help planners trust the recommendations coming from the IO tool. This is of critical importance as more calculations are informed by machine learning capabilities that continually test and challenge the underlying assumptions planners have had. Vendors that have been able to make the shift to cloud are better positioned to leverage the analytic capabilities that will drive value for inventory optimization solutions in the years to come.

· · · · ·

4R Systems

4R Systems remains a Core Provider in the 2018 IO Value Matrix, delivering a cloudbased inventory solution for retail customers. With the goal of maximizing profitability, 4R Systems has capabilities that include forecasting, seasonal allocation, vendor order optimization, and assortment optimization. The vendor has been focusing on enhancements to its replenishment capabilities, applying machine learning to predict where customers might have issues with their replenishment plans.

4R Systems delivers additional machine learning capabilities that test a customer's historical understanding of settings such as thresholds. The software analyzes changes at various levels of inventory such as categories and SKUs and then feeds the changes back into the threshold definitions for use in creating future forecasts and demand. The system develops overrides to improve and hone the definitions set by the user.

Currently, 4R Systems has a service team that designs the solution for customers to get optimal results and allows the user to see the financial impact of a deviation from the original plan. The vendor uses a suite of simulations in its service delivery. It is looking to give users more control over the simulations the software provides. As a SaaS offering, 4R Systems can be deployed quickly and can scale as the business needs evolve. Moving forward, additional investments in capabilities such as space optimization should expand the functionality of the solution.

Blue Ridge

Blue Ridge moves into the Facilitator Quadrant in the 2018 IO Value Matrix. As a cloud-native vendor, Blue Ridge focuses predominately on helping customers manage finished goods in wholesale distribution and retail. Leveraging a single data model, Blue Ridge delivers all its capabilities through one solution rather than integrated modules. With a suite of forecasting methods, item modeling, and product lifecycle management, the vendor has automated steps of the inventory optimization process which allows customers to set stocking policies and leave the solution to manage the activation, purchase, and deactivation of items based on the prescribed parameters.

A significant focus of Blue Ridge has been to deliver enterprise-grade capabilities through applications that resemble those on mobile devices, with simplicity and usability being central to its development efforts. Blue Ridge has also been enabling customers to perform continuous end-to-end planning, which requires prescriptive analytics connected to financial objectives and visibility across the supplier, distribution center, and storefront. Deployed in the public cloud on Amazon Web Service, Blue Ridge is bringing greater flexibility to its users, allowing them to perform planning tasks and reconfigurations through any web browser. Continued investment in functionality that provides automation and prescriptive action recommendations will help Blue Ridge deliver additional value and improve their position in future editions of the Value Matrix.

E2open

E2open continues as a Leader in the 2018 edition of the Inventory Optimization Value Matrix, offering a cloud-based multi-enterprise inventory optimization (MIO) as a piece of its overall capabilities which includes Sales & Operations Planning, Supply Planning & Response, Demand Planning & Sensing, and supplier and channel management across multiple tiers. Consistent with its end-to-end solution, E2open's MIO solution sits between demand planning and supply planning, pulling in data from the enterprise resource planning (ERP) system, channel management solution, and demand planning & sensing, helping companies set optimal inventory and safety stock levels that are fed into the supply planning and response module as well as further upstream to suppliers.

The value proposition for the MIO capabilities rests on its roles as a core planning component of the integrated solution, offering closed-loop analysis of stocking levels by monitoring execution of inventory targets and collecting data such as lead times, SKU-location interdependencies, and policy violations. The tool shows inventory at customer-facing and interplant locations with the demand planning forecasting using demand propagated throughout the value chain, eliminating the need to pool risk and resulting in more accurate supply plans allowing customers to cut inventory holding levels.

E2open has made several acquisitions that have added to its footprint and helped it address its customers' needs, including Orchestro for point-of-sale data collection, Zyme for channel data management, Terra Technology for demand sensing, and Steelwedge for sales and operations planning (S&OP) (Nucleus Research, R42—*E2open merges with Steelwedge*, February 2017; Nucleus Research, Q41—*E2open acquires Terra Technology*, March 2016). Last year the vendor debuted its new user interface which ties together the acquisitions it is integrating into a seamless experience. As a result, users progress through a planning cycle without switching between applications and data moving between modules. With a focus on extending inventory optimization efforts to include upstream supply networks and downstream demand channels, E2open gives its customers the chance to perform a single solve across their entire value chain.

GAINSystems

GAINSystems continues as a Leader in the 2018 edition of the IO Value Matrix, offering an integrated multi-echelon inventory optimization (MEIO) solution, along with demand planning and forecasting, replenishment and production optimization, and sales, inventory, and operations planning (SI&OP). Delivered on-premises or in the cloud, GAINSystems' forecasting engine draws from 40 statistical models to automatically select the most plausible demand model to use when building a forecast focusing primarily on distribution, manufacturing, and maintenance, repair, and operations (MRO).

Forecasting can be performed from the bottom-up, starting with granular SKUlocation or channel-level data, that can then be aggregated based on SKU attributes to ensure the forecasts preserve the characteristics of the items. GAINSystems can also determine the most profitable distribution options for each SKU across the customer's network. Customers use GAINSystems' inventory optimization capabilities to determine the optimal replenishment quantity and service stock parameters to minimize total annual costs. The optimal stocking level can be made at the SKU-location level while incorporating comprehensive errors from supply and demand variability, which analyzes historical data to project uncertainty tolerances across cycle count error, forecast error, lead time uncertainty, and demand variability.

On the automation front, GAINSystems has a few applications that seek to eliminate manual processes through supervised machine learning. In addition to enabling customers to use no-touch purchasing with automated replenishment planning, GAINSystems is automating forecast approvals, master data management, and demand outlier detection. GAINSystems has also been investing in UI improvements that require no coding to configure. The system has 20 default visualizations that can all be tailored to the user needs. Additional UI investments should help GAINSystems' positioning in future editions of the IO Value Matrix.

Infor

Infor continues as an Expert in the 2018 IO Value Matrix, having kept pace with the market as it integrates capabilities from acquisitions it has made and continues its push toward the cloud. Offering a networked supply chain, Infor has capabilities that span supply chain planning, execution, finance, and procurement, all underpinned by a multi-enterprise business network, leveraging capabilities it acquired when it purchased GT Nexus in 2015. Having migrated many of its legacy capabilities to its cloud offering, Infor is positioned to deliver faster time to value for its customers.

Within supply chain planning, Infor has modules that cover demand planning, supply planning, production planning & scheduling, and S&OP. The vendor's IO capabilities start with segmentation and include simulations, stock rebalancing, and costoptimizing. Infor offers visibility outside the enterprise with its multi-enterprise network as well as distributed order management to help with orchestration and fulfillment. Infor has been investing in its usability, adding more workbenches and visualizations to help planners be more efficient, such as side-by-side inventory simulation comparisons. Additional investments in cloud verticals are paying dividends for customers who have more of their industry-specific needs met out-ofthe-box. Continued acceleration of its cloud business should see customers achieving value more quickly and improve Infor's position in future editions of the IO Value Matrix.

JDA Software

JDA is a Leader in the 2018 Inventory Optimization Value Matrix, based on the value customers realize using its Inventory Planning module, which can be delivered as a point solution or as part of JDA's end-to-end planning solution suite. Inventory Planning shares the same schema and data model as JDA's other modules, which cover network design, S&OP, demand planning, master planning, replenishment planning, factory planning, scheduling, and order promising.

In Inventory Planning, users can perform inventory reviews at multiple levels, from strategic to tactical to performance. In addition to traditional forecasting methods, JDA is leveraging machine learning techniques to detect planner bias as well as feeding actual demand levels to an algorithm that generates stocking level parameters. Additionally, JDA customers can use machine learning to predict service failures and offer prescriptive remedies.

On the roadmap, JDA is looking to introduce inventory planning algorithms for slow moving and lumpy items as well as visualization and root cause analysis usability enhancements. Looking past 2018, JDA is working on furthering its segmentation capabilities with an algorithm that automatically clusters items based on their attributes as well as holistic inventory optimization throughout the entire network to facilitate a single solve across multiple trading partners and stakeholders.

JDA has partnered with Google to leverage its Cloud Platform capabilities, on which JDA is delivering its software as a service (SaaS) applications. The partnership is moving JDA towards more cognitive demand planning, which will take advantage of real-time demand sensing, impact analysis of external influences, and digest large streams of data such as Internet of Things (IoT) signals. Although it is too early to tell how much extra value this will deliver for customers, JDA's investments in machine learning are proving beneficial and driving better performance for those customers that have implemented them.

Kinaxis

Kinaxis is a Facilitator in the 2018 edition of the IO Value Matrix. The Canada-based vendor focuses on supply chain planning, offering 13 different applications that make up its RapidResponse solution. Kinaxis' inventory planning and optimization (IPO) application can be delivered as a stand-alone point solution, but in general, customers use the optimization engine in conjunction with other applications delivered the Kinaxis platform, thereby deriving better value. For example, users can generate the demand forecast with the demand planning application, where the tool automatically selects the forecasting method that best fits the demand pattern of each SKU. The forecast is then brought into the IPO application, where users can optimize inventory levels to meet projected demand. Kinaxis leverages a single data model across all its applications, which helps users move more seamlessly from one application to another as they go through their planning processes.

Since the last iteration of the IO Value Matrix, Kinaxis has delivered some of the usability enhancements that were on its roadmap in 2017. In its August 2017 update, Kinaxis added new business process flow building capabilities to help users set up and start using the application faster, speeding their time to realize value. Additional usability improvements allow users to set up rules-based processing and product segmentation more easily, with the same goal of helping users get up and running more quickly. From a functionality standpoint, Kinaxis released several improvements to helping users cleanse and prepare their data. New tools that detect and correct outliers and analyze data integrity were also included in the August 2017 update.

On the roadmap for 2018 and beyond, Kinaxis is planning to make usability improvements such as network visualization capabilities within IPO. Additional roadmap elements include the ability to add group-level service levels, thereby helping users segment and aggregate by service requirement as well as cost modeling that quantifies the entire product lifecycle, so users can review total item costs and associated trade-offs. Finally, Kinaxis is looking for practical ways of applying machine learning capabilities such as investigating future product lead times based on analysis of historical data. The vendor is also developing an analytic tool that looks for potential risks in the inventory optimization plan, so users can decide what level of risk they are willing to accept.

Focused squarely on the planning side of supply chain management, Kinaxis relies on its integration capabilities to provide the feedback loop from a customer's execution system, such as an enterprise resource planning solution. Kinaxis' continued investments in helping user adoption should pay dividends by ensuring that customers realize value from the IPO application quickly. Additional investments in automation capabilities will help Kinaxis' positioning in future editions of the Value Matrix.

LLamasoft

LLamasoft is a Facilitator in the 2018 IO Value Matrix, focusing on supply chain network design that enables customers to make tactical decisions and optimize inventory. LLamasoft provides a suite of technologies that help customers in addition to network design, including MEIO, service level optimization, inventory simulation, and scenario analysis. The vendor has vertical capabilities covering manufacturing, retail/consumer and packaged goods, healthcare/pharmaceuticals, and chemicals.

Since the last Value Matrix, LLamasoft has released capabilities that allow customers to build disposable applications. Customers can configure thresholds and automated alerts with the applications to facilitate a self-healing supply chain, driving value through greater user productivity. LLamasoft is also deploying machine learning in its demand modeling to analyze predicted lead times and prompt re-forecasts when the values change. Customers realize value from the solution by evaluating network changes for alternative strategies that can lower costs and improve efficiency. LLamasoft's position in the Value Matrix should improve as more customers start to use the application to build functionality.

Logility

Logility is a Facilitator in the 2018 IO Value Matrix, delivering the market standard for usability. With its Voyager Solutions, Logility has product capabilities that span supply and demand optimization, replenishment planning, manufacturing planning, assortment and allocation planning, and transportation and shipping optimization. Offering customer-centric continuous planning, the vendor delivers a platform that enables companies to integrate with suppliers, collaborate with their customers, and leverage analytics that draw on data from outside the enterprise including social sensing and unstructured information.

The vendor helps customers with their master data management using capabilities delivered by Voyager AdapLink, which it deploys as the integration layer between the supply chain and the enterprise's data. In addition to offering out-of-the-box templates for connecting with ERP systems and reducing data lags between systems, the integration allows for multi-enterprise collaboration. Logility has also been investing in its analytics capabilities that are embedded in Voyager Solutions. Drawing on external sources such as syndicated market data and specialized data for verticals, Logility is applying machine learning to deliver automated simulations rather than manually created scenarios. A focus for the vendor moving forward is integrating its latest acquisition, Halo Business Intelligence. We expect Logility to push software usability forward, as well as deepen its analytics functionality, as it continues to invest in bringing Halo's capabilities to its core products.

Manhattan Associates

Manhattan Associates is an Expert in the 2018 Inventory Optimization Value Matrix, delivering a suite of planning tools that enable its omnichannel inventory optimization (OIO) capabilities. With its solution, Manhattan Active Inventory, Manhattan Associates aims to link customer fulfillment experience initiatives with inventory optimization strategy, accounting for the changing requirements of companies—particularly in retail—that are focused on offering multiple fulfillment channels to their customers. The vendor wraps inventory optimization around modules addressing S&OP, assortment planning, promotional planning, demand forecasting, replenishment, vendor managed inventory, and multi-echelon inventory.

In addition to investing in OIO, Manhattan Associates is bringing analytics to its solution that helps planners evaluate the impact of promotions and maintain visibility of the larger inventory picture. With promotional event analysis, the vendor is leveraging machine learning to identify the extent to which promotions shape demand, providing feedback to marketing departments. The capability also considers third-party data to add context to the demand signals, such as weather data.

To improve end-user usability, Manhattan Associates has also embedded intuitive data visualizations directly into commonly used workflows. For example, visualizations added to the suggested order approval workflow dramatically reduces the time spent reviewing orders. For executives and supply chain leaders, they've also added dynamic dashboards to ensure that financial performance from inventory is constantly aligned with overall enterprise strategy. While its investments are encouraging, Manhattan Associates' path to the cloud has been slow, which impacts how quickly customers have been able to achieve value. As its cloud business accelerates, we expect Manhattan's value proposition to improve.

One Network

One Network is a Leader in the 2018 IO Value Matrix, with customers realizing value from the multi-enterprise network capabilities delivered by the vendor's real time value network (RTVN). Connecting planning with execution, One Network leverages each execution cycle as an opportunity to tune the planning software, which includes control tower, demand and supply planning, forecasting, and simulation in addition to inventory planning and optimization.

Drawing its model directly from execution and transactional data derived from connected downstream trading partners, One Network customers can forecast with real lead times and variability, rather than assumed values. One Network also deploys learning agents that analyze transaction data to discern demand patterns that are fed into the forecast. The solution also runs micro-simulations to evaluate the targeted inventory levels against the end-to-end execution details. The simulations are analyzed with a machine learning algorithm to evaluate the outcomes from each planning cycle with the outcomes.

One of One Network's stand-out strengths is its automation capabilities which allow customers to operate parts of their supply chain with little to no user intervention. With AI deployed to capture demand sensing as well as execution details, One Network enables customers to automate replenishment plans based on visibility of demand and on-hand inventory within the extended value chain in conjunction with real lead times and logistics information. Based on the business rules established by the user, the software can independently manage and fine-tune the safety stock levels, leaving planners to address exceptions to the exceptions. Continued investment in automation and low-touch capabilities will see One Network's place in the Value Matrix improve further.

Oracle

Oracle is a Facilitator in the 2018 edition of the IO Value Matrix, offering a suite of value chain planning tools. Though predominately installed on premises, Oracle has made significant investments to build its Supply Chain Planning (SCP) Cloud solution from the ground up. It has also taken best practices from its cloud user interface back to its on-premises deployments to increase the usability of the product. Within SCP Cloud, Oracle has simplified the solution landscape, so the users see only the functionality they need rather than switching between modules. Offering its capabilities as cloud-based services rather than individual applications, Oracle is focusing on ensuring that its capabilities are fit-for-purpose to increase the value customers realize.

Since the last Value Matrix, Oracle has continued to make its user experience a highlight of its product, offering contextually driven charts, graphs, and visualizations that help users move through the process with a better understanding of where they stand on KPIs of their choosing. As Oracle's cloud business continues to accelerate, its ability to deliver value will improve. In the meantime, its steps to help customers transition to the cloud via hybrid deployments and modernized UI ensure it is keeping pace with the market.

SAP

SAP is an Expert in the 2018 edition of the IO Value Matrix. SAP delivers IO capabilities as part of its Integrated Business Planning (IBP) solution, which covers five planning modules: S&OP, inventory, demand analytics and sensing, control tower, and supply planning and response. Available on the HANA cloud platform, SAP IBP is focused on end-to-end supply chain management, with customers realizing greater benefits when they have more than just the inventory optimization module and can feed forecasts and demand sensing data into the optimization engine.

Since the last edition of the Value Matrix, SAP has focused on bringing usability and algorithmic enhancements to the IO product. Scenario planning and side-by-side what-if analysis are driving better usability for customers, who already use the Fiori user interface. As part of its continuous efforts to improve its forecasting and optimization algorithms, SAP is looking to enable demand-driven material requirements planning (DDMRP) to the solution suite as well as additional machine learning capabilities.

Continued investment by SAP in increasing the level of automation and machine-driven insights that customers can leverage will help SAP deliver better value. SAP has made strides to ease the implementation complexity for customers by offering a cloud deployment option, further investment in helping customers get established and operating quickly will accelerate customers' time to value. As the solution matures, customers will be able to take advantage of the larger SAP ecosystem of partners and user community to extend the value they get from the solution.

Slimstock

Slimstock is a Core Provider in the 2018 IO Value Matrix, offering its Slim4 product which includes demand profiling, forecasting, demand planning, S&OP, replenishment planning, and MEIO planning. Slimstock services the needs of customers from many industries including automotive, retail, wholesale, manufacturing, spare parts, industrial, consumer packaged goods, construction, and healthcare.

Since the last Value Matrix, the vendor has been investing in updating its UI to be web browser-based as well as in a performance monitor application that is configurable to show the user the most relevant KPIs. Operating as a knowledge partner with its customers, Slimstock offers fixed-fee implementations and customers generally go live in three to four months. The vendor is positioning itself to deliver more SaaS in the future and already offers a plug-and-play connector with ERP systems. The investments Slimstock is making in updating its UI should improve its position in the IO Value Matrix in the future.

Smart Software

Smart Software is a Core Provider in the 2018 IO Value Matrix, with its Smart Inventory Optimization (SIO) solution offering inventory planning policy decision support and impact tracking. Users can provide their own planning parameters or allow the software's optimization logic to prescribe planning parameters and service levels. Delivered through web-based UI, Smart Software's cloud products include demand planning and forecasting in addition to inventory optimization.

Since the last Value Matrix, Smart Software announced plans to deepen its partnership with Epicor, an ERP software vendor. The goal of the partnership is to integrate Smart Software's inventory planning and optimization capabilities into Epicor's ERP solution. The integration is designed to better link operational planning with execution. However, additional investment is required for the vendor to keep pace with the technologies being deployed in the market by other vendors.

ToolsGroup

ToolsGroup continues as a Leader in the 2018 edition of the Inventory Optimization Value Matrix after making strides to productize its machine learning capabilities into a suite of applications that address specific situations within a customer's supply chain, such as new product introductions and seasonality clustering. Delivered as part of its Service Optimizer 99+ (SO99+) solution, ToolsGroup's inventory optimization capabilities can be delivered as a stand-alone offering, but partners with the vendor's demand planning, supply and demand collaboration, production planning, sales, inventory, and operations planning (SI&OP), and demand sensing capabilities on a holistic platform. Sold predominately as a SaaS solution, ToolsGroup leverages a single data model that ensures the distribution model is preserved throughout the entire planning solution, so information is not lost as the planner moves from demand planning to inventory optimization to supply planning.

An area of focus for the vendor has been around its machine learning engines which can support several scenarios. Some of the capabilities draw from non-traditional inputs and data streams such as weather, web sentiment, IoT data, and customer segmentations. The applications cover advance forecasting, early signals, events forecasting, events auto-detection, new product introduction, and seasonality clustering.

To aid usability, ToolsGroup uses several visualization engines including Microsoft Power BI, delivering data to the user that is easily digestible and flexible, as well as ePlanner which helps users manage events and promotions. ToolsGroup also uses Microsoft Azure Cloud services to store data for use in machine learning scenarios as well as delivering the compute to power its simulation capabilities. With the strides it has made in bringing insights to users with machine learning products, ToolsGroup is making it easier for customers to derive value from the solution and realize better forecasting and inventory optimization results.

Vanguard

Vanguard moves into the Leaders quadrant in the 2018 edition of the

Inventory Optimization Value Matrix, with customers leveraging the MEIO capabilities as part of its Forecast Server product line. In addition to inventory optimization, Forecast Server delivers sales forecasting, demand planning, S&OP, integrated business planning, and financial planning and analysis. Customers derive value from Vanguard's analytics capabilities, which draw on a suite of forecasting models and AI-assisted automatic selection, as well as the flexibility of the system to match a variety of industries and customer use-cases.

With a focus on reducing forecast error to minimize safety stock, Vanguard draws on 33 time-series forecasting methods to address scenarios such as seasonality, business cycles, and demand trends. Customers can run simulations based on discrete events or use Monte Carlo methods to dynamically measure the potential impact on key performance indicators (KPIs). Deployed on-premises or in the cloud, Vanguard's platform delivers a single data model across all its modules, which ensures that planners see the same version of the truth. Additionally, Vanguard has invested in the power of its platform to manage up to 16 million SKUs, ensuring that customers with complex product-mixes can plan down to the most granular level.

Delivering a web-based user interface, Vanguard supports mobile users and allows for multi-enterprise network capabilities where users can integrate suppliers and trading partners into the planning process. The flexibility of the platform helps with user adoption, as Vanguard works closely with customers to ensure that the workflow is correctly configured to the use-case. Vanguard is also providing automation capabilities based on business rules that are managed by an AI engine. Additional investments in automation will see Vanguard's position in the Value Matrix continue to improve.

Content Management Value Matrix

Dot indicates current position. Arrow indicates trend in 2019 relative to others in the market.

CHAPTER 9
CONTENT MANAGEMENT

Those vendors with a clear roadmap for improving the user experience, process automation, seamless integrations, and robust security are differentiating themselves in the market. In 2017, the CM mantra was "digitization and moving to the cloud." 2018 continues along the same trajectory, but with a greater focus on the user experience. Customers expect solutions that are flexible and agile, and that can solve evolving needs while adding value to their business. Some trends identified since last year are:

Data migration will become obsolete in the next five years: File migration is the traditional method of reaching data for editing, storage, collaboration, and integration with other business processes. However, file migrations can be time-consuming and inefficient. They can also test the data storage limits of an enterprise. Metadata is a data set that gives information about other data. It is a resource for data discovery, content classification, and contextual search, enabling users to access information more easily. Repository-agnostic access is the ability of a solution to manage content from any database, regardless of location. Uniting metadata with repository-agnosticism allows the user to work with data and content at its source, eliminating the need for costly and time-consuming data migrations.

Automation and robotics: AI and ML advances for process and work-flow automation will impact return on investment (ROI) with greater productivity, increased efficiency, and improved risk management due to the reduction of human errors. Although automation is not a new concept, the exponential growth of data combined with technological advances such as

Robotic Process Automation (RPA) are growing in acceptance, driving additional value to the client.

The hybrid solution is back: Vendors remain committed to digitization and cloud services but have recognized the value of a phased transition with solutions that function both on-premise and in the cloud. Although newer vendors tend to offer cloud solutions, vendors were having difficulty transitioning existing users from on-premise to the cloud. In addition to the challenge of change-management issues, some users have a significant investment in hardware that would be difficult to justify dismantling, while others have regulatory requirements for data geolocation and storage. Consequently, more vendors are offering hybrid solutions as a transitional option for clients as they work on moving to the cloud.

· · · · ·

Alfresco

Alfresco provides a single-platform approach to managing content, digital business processes, and governance through the Alfresco Digital Business Platform. Customers can also purchase the Content Services and Process Services components of the solution separately. Additionally, Alfresco offers functionality for AI, analytics, integrations, and an open API Application Development Framework that allows for integrations with other business intelligence and visualization tools. Alfresco provides metadata capabilities that allow for intelligent searching. The vendor focuses on a cloud-centric strategy, though it also provides on-premise, hybrid, and managed solutions.

Like other CM vendors, Alfresco is committed to digital transformation, leveraging technology to improve performance and agility. The Alfresco roadmap is the development of a single, modular platform for digital business including content management, business processing, and governance with modern open architecture. Alfresco is working to improve usability and visualizations, and is also working to reduce time to value with custom-focused applications. Additionally, the use of common components for both content and process applications will foster an open ecosystem. Developing partnerships are becoming more important, an example being Alfresco Quick Start on AWS, an CM configuration that is adaptable for multiple scenarios with out-of-the-box deployments.

Box

Box offers a single platform solution that provides cloud content management, collaboration, business processing, security, and regulatory compliance. The solution provides a full range of functionality and focuses on providing an intuitive end-user experience that can be tailored to the specific needs of a client. Box has a strong partner network including IBM, Microsoft, Adobe, AWS, DocuSign, and Salesforce that enables it to provide additional functionality to end-users. Box provides many pre-built integrations that connect its platform to third-party applications, allowing end-users to bring outside content into Box. An application programming interface (API) layer surrounding the platform also empowers users to build custom integrations with other applications. The solution provides workflow management through integrations with Nintex, Pega, and IBM Case Manager. Box Relay lets end-users create custom, automated workflows.

This past year has been a transition for Box into the complex CM market. End-users interviewed were almost unanimous in their praise for Box's innovation and commitment to further platform development, rating the Box support teams as well prepared and willing to spend whatever time necessary to resolve an issue. Box plans to expand on the size and scope of the ecosystem using AI to enhance auto-classification, categorization, records management, and development of a "threat detection" capability for increased security and data protection. As Box continues to develop functionality, we believe it will be a notable CM provider.

Comarch

Comarch offers an CM solution that addresses a full range of content management needs, including document capture, optical character recognition (OCR), workflow, records, lifecycle, and archiving. It also provides capabilities for master data management and business to business data exchange. Its content management services serve as the base for its packaged workflow automation solutions, including solutions for accounts payable, accounts receivable, e-invoicing, procure-to-pay, and contract management. These solutions are designed for and offered to specific verticals including retail, manufacturing, healthcare, banking, and insurance. Comarch also allows for integrations with a variety of ERP, CRM, and HR platforms.

Comarch is focused on a strategy of organic growth. Comarch is concentrating on research and development for the near future to broaden capabilities as it expands into the North American market. As a Europe-based company, security and compliance are key strengths for the company. Comarch has a clear roadmap and long-term strategy to add functionality and business services to its current offerings. Comarch is committed to research and development in its promise to expand functionality. We regard Comarch as a solid core vendor with the potential to grow in the North American market.

Digitech Systems

Digitech Systems offers a full suite of software and cloud services that provides an end-to-end system of automated data capture, content management and collaboration, secure content storage, and automated business processes. The solution enables users to efficiently capture information as it enters a business and then to automate the flow of that information through their business. As a result, users can access insights at the right time to make informed business decisions. Digitech distinguishes itself from other vendors because it uses proprietary AI algorithms. Digitech's patented AI software builds a multi-dimensional profile of all captured content, and ML continually increases the accuracy of the system.

Digitech continues to invest in the development of its solution. It continues to offer enhanced product launches such as ImageSilo. It is also moving to object-based storage, which will eliminate the need to identify a file structure to locate a specific document or information. Digitech is unique in that it owns its proprietary capture and storage process, having these capabilities from the ground up. With ongoing product development, strong usability, robust functionality, and a loyal user base, Digitech may be one of the strongest vendors that most end-users have never heard of.

Docuware

DocuWare has a strong presence among mid-market companies in the United States, Germany, and France with sustained growth of on-premise customers and accelerated growth in cloud-based customers. It also sees success in departmental deployments in large enterprises. The solution offers strong out-of-the-box functionality for document management and workflow automation without the heaviness of the largest legacy CM vendors.

DocuWare competes with smaller and more local CM cloud players but often wins deals because of its global presence and its high usability due to an intuitive drag-and-drop UI.

DocuWare is committed to a zero-interruption policy to eliminate operational downtime and maintain stability. The company plans to enhance usability in 2018 with pre-configured solutions and templates, in response to client feedback. Strong security features with document and communication encryption support regulatory and records policy compliance. Workflow automation improvements will continue as part of the roadmap DocuWare has identified for the future.

Egnyte

Egnyte has significantly enhanced its offering in the past year, transforming from a primarily EFSS solution to a full CM solution with BPM functionality, AI and ML, GDPR compliance, and strategic partnerships. Recent additions to functionality such as custom search ability with metadata, simplified cloud migration, enhanced security, and strategic partnerships place Egnyte firmly in the content management arena. Egnyte has customers around the world and serves a substantial number of large enterprise customers.

Egnyte plans to continue its progression to full-service CM with a unified platform for comprehensive content management. Offering on-premise, hybrid, and cloud deployments with on-going infrastructure modernization and content-centric solutions, Egnyte is a progressive company with the potential for greater market penetration.

Epicor Docstar

DocStar offers CM and BPM with strong core functionality and prebuilt integrations. Since being acquired by Epicor in January 2017, DocStar has extended its integration portfolio to include Epicor ERP solutions. DocStar recently improved the usability of its forms, enhanced its capture process, improved its search capabilities, and expanded its automated workflow functionality including the addition of an end-to-end case management tool. DocStar has also leveraged the Epicor capabilities in areas such as business analytics and mobile platforms.

DocStar is committed to the Epicor "cloud first" strategy but continues to be available for on-premise implementations as well. DocStar has plans to offer e-signatures capability, enhanced contract management, Sales Order Automation, and further expansions of its capture technology in 2018.

Fabasoft

Fabasoft is based in Austria and offers CM and BPM to large enterprise customers primarily located in Europe. The Fabasoft Cloud Platform provides capabilities for modeling, project management, test automation, encryption, and big data search. Specifically, the platform delivers functionality for automated workflow processes including digital personnel file management, contract management, digital asset management, board communications, collaboration, process management, and compliance management. Fabasoft also provides vertical solutions tailored to manufacturing, healthcare, financial institutions, and the public sector.

Fabasoft recognized that its focus on improving usability in 2017 needed to pivot back to functionality and is preparing for the future using AI and machine learning for intelligent document scanning, management, and indexing. E-mail inbox management is available for unstructured documents. Fabasoft continues to invest in expanding the functionality of the systems to improve client models and offerings.

Hyland

OnBase by Hyland provides an approach to document management that focuses on managing content, processes, and cases within purpose-built applications that address vertical-specific needs, while still providing a centralized hub of information that any user can access via their primary line of business application. Hyland provides particularly robust solutions for financial services, insurance, government, higher education, and healthcare. The OnBase platform offers solutions for invoice processing, contract lifecycle management, and HR services. Hyland also offers extensive integrations to connect with other business applications. Additionally, Hyland provides an intuitive user experience. WorkView | Case Manager lets administrators build data driven applications and workflows with native, low-code configuration tools.

OnBase plans to continue its road-map for developing and enhancing application-centric solutions and process automation and workflow technology. OnBase has committed significant financial resources to accomplish these goals, including improved usability, configurable reporting dashboards, capture with Brainware by Hyland, case management, and e-signature functionality. OnBase is actively modernizing its entire platform and expanding its content services offerings.

IBM

IBM offers an extensive content management portfolio, the foundation of which is made up of Content Navigator, FileNet, and Datacap. At the core of all these content services are advanced document capture, metadata tagging, advanced search, and collaboration capabilities, with ML, NLP, and analytics capabilities throughout the platform. These content services seamlessly integrate with the DBA Platform to facilitate efficient workflows. IBM's Content Navigator enables end-users to search, preview, access, and edit content in-place from almost any on-premise or cloud repository and has an extensible framework so that users do not have to migrate content. In the past year, IBM has made user experience updates to Content Navigator and integrated Content Navigator with Office 365. The new FileNet Cloud Storage 5.5 was released in December 2017, and provides end-users with the ability to store and collaborate on content. Datacap has multichannel capture capabilities, extracting document content for automatic classifications with the appropriate metadata. IBM's newest release, Datacap 9.1.3, enhances the tool's high-speed scanning, bulk document processing, and NLP capabilities, with a connector to Watson visual recognition. Meanwhile, Datacap Insight Edition provides cognitive capture capabilities enhanced by ML, NLP, and advanced imaging classification.

IBM is an established company with multiple offerings and a considerable legacy user base. IBM remains ahead of the curve of relative to other legacy vendors in its movement to the cloud. Its content management solutions support complex processes across enterprise operations and address almost every imaginable need a corporation can define. Though the size of an on-premise IBM deployment can be a deterrent for organizations who do not have a significant amount of time or resources to devote to such a project, the IBM Cloud offerings have increased in popularity amongst customers who do not want to make a big upfront investment. We expect that the IBM DBA Platform initiative will have a positive impact on the CM space. Robotics and work-flow automation should address the universal problem of repetitive and time-consuming manual CM tasks. Roll-outs of additional automation and straight-thru-processing across an enterprise are planned for 2018 and are expected to have a positive impact on the customer experience with an end-to-end work-flow solution. Enhancements that should affect CM include cognitive capture, low-code modifications, operational intelligence, and the end-to-end cloud solution.

iManage

iManage is a vertical specialist with a platform optimized for the legal industry, accounting firms, financial services, and professional services with the iManage Work 10 suite. Its solution offers core document management, records management, and document-centric collaboration capabilities. It has identified "content" as the core work product of law firms, accounting firms, financial institutions, and professional services. iManage provides artificial intelligence and an easy-to-navigate UI that enables users to classify content, extract information, and identify relationships between content through a knowledge graph. As such, the goal of iManage is to enable users to "shop for content" as easily as one shops on Amazon. Additionally, iManage Share provides secure external file sharing and collaboration.

Most enterprise law firms that use iManage remain on-premise and will likely remain so. We anticipate that iManage will continue to be a strong provider in the legal and professional services spaces and will expand its functionality and grow its customer base over the next year.

Laserfiche

Laserfiche offers highly usable, industry-specific solutions for content management and BPM for mid-size to enterprise customers. The vendor's industry-specific solutions in its BPL are developed organically with its user community and global resellers. In government, for example, Laserfiche Workflow reduces costs, enables transparency, and streamlines processes for Freedom of Information Act requests. In financial services, Laserfiche Forms automates task management for banks, credit unions, insurance companies, and wealth management firms with scalable solutions. In education, the Laserfiche process automation suite modernizes campus operations and streamlines communications with students, faculty, and staff. For business processing and supply chain management, Laserfiche enables users to automate invoice and accounts payable processing. With Laserfiche's healthcare workflow automation, users can link insurance information and lab reports to electronic medical records and streamline bills and claims processing. Additionally, Laserfiche's justice systems solution allows officers to access records on mobile and search full texts of witness statements and depositions.

Laserfiche has a clear, client-centric roadmap as an overall content management eco-system. Records management, security, compliance, and task and process automation are the focus for the near future. AI and ML are featured components in the roadmap, which leverages these technologies to deliver a comprehensive suite of services. Some of the newer features include geolocation and business rules. Robotic Process Automation (RPA) is also in the pipeline to reduce daily, repetitive manual tasks through rule-based automation.

M-Files

M-Files enables end-users to tag documents and non-document objects with metadata and then categorize and process information based on what it is, not where it is. Content can be tagged with almost any information important for understanding it in a business context, including case, claim, asset, contact, employee, and so on. M-Files uses AI to facilitate document classifications by giving metadata tag suggestions to end-users. Users can also automatically restrict access permissions on content by tagging content according to its associated customer and project. Once tagged, users can search through information using various metadata tags and manage information in multiple ways based on metadata. As a result, the same information can be accessed and used differently by different users or groups based on context, without duplication.

M-Files plans to continue to develop additional automation, launch new features more quickly, enhance its portals, and provide more frequent cloud updates. With continued progression and a differentiated approach, we believe M-Files will continue to deliver.

Micro Focus

Micro Focus Content Manager is a governance-based application designed for highly regulated enterprise businesses. Its unified solution supports a wide variety of content types and can meet regulatory and governance mandates for many jurisdictions worldwide. Micro Focus provides an approach known as Secure Content Management (SCM), which focuses on a data management strategy that takes a proactive approach to managing risk and protecting information through a broad and integrated portfolio that secures and governs information. The SCM approach is holistic and focuses on controlling data, keeping data secure, data archiving, file analysis, intelligent classification, and the interconnectivity between file analysis, structured data archiving, and content management. Advanced analytics throughout the system provides automated classification and policy enforcement.

Micro Focus continues to execute on its core strategy of information governance using AI to extend automation and reduce error rates occurring with a manual task. In addition to content management, Micro Focus creates business value with improved productivity and reduced IT costs.

 ## ROI Case Study: Micro Focus

Wandera • ROI: 270%

Wandera deployed the Vertica analytics database management platform to replace an AWS Redshift legacy system and to improve the performance, reliability, and scalability of its mobile protection products. The implementation of the Vertica platform enabled Wandera to avoid hiring additional technical staff to support the legacy platform and improving the IT team's (managing the data platform) productivity by 18 percent. Additionally, Wandera experienced better performance and reliability with Vertica, reducing hardware issues by 5 percent, and achieving an average annual benefit of $339,792.

THE COMPANY

Wandera provides enterprise mobile security and data management solutions, focusing on prevention, detection, and containment of data issues. Its software as a service (SaaS) platform provides customers with a Secure Mobile Gateway (SMG) that compresses mobile data, enforces acceptable usage policies, provides multi-level mobile threat detection, and provides data usage reporting. Wandera solutions are designed to secure mobile devices, optimize data usage, and protect sensitive business data.

CUMULATIVE NET BENEFIT

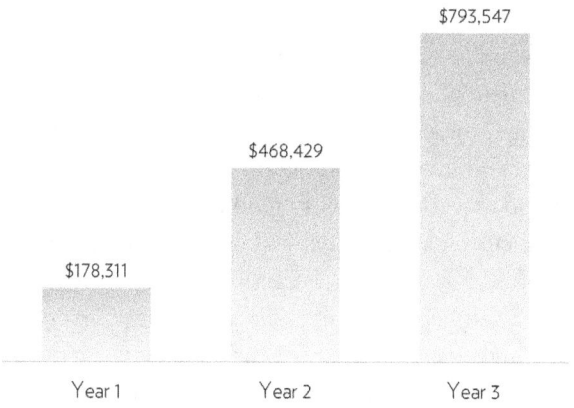

$178,311 Year 1
$468,429 Year 2
$793,547 Year 3

KEY BENEFIT AREAS

Moving to Vertica enabled Wandera to replace a legacy system, improve performance, reliability, and scalability of its mobile protection products while reducing costs and increasing productivity. Key benefits of the deployment include:

- **Cost Savings.** With the deployment of Vertica, Wandera was able to avoid hiring additional IT staff that would be needed to support the legacy system. Additional savings were achieved with Vertica's lower costs for data storage and elimination of hardware.
- **Improved IT productivity.** Wandera can support the Vertica platform with minimal resources freeing-up as much as 30 percent of IT time for more strategic activities.
- **Increased user productivity.** Moving from Redshift to Vertica to manage its data, the staff accomplishes more, in less time using Vertica. Additionally, the improved processing speed and reliability of Vertica ensured that real-time access to data is consistently delivered.

- **Improved technology management.** Eliminating Redshift enabled the company to save almost $400,000 of direct expenses while delivering a more reliable system.

BEST PRACTICES

With Wandera's choice to leverage the cloud for data management and storage, it avoided investing in hardware and annual maintenance costs required by the previous platform and increased productivity and agility. Wandera did opine that if they were to redo the deployment, they would focus on a tightly defined scope, rather than the broad offerings, to avoid distractions. Using Vertica, Wandera can deploy big-data management to its customers quickly and cost-effectively. In turn, its customers will benefit from a lower TCO and the ability to access information rapidly and improve company performance.

· · · · ·

Microsoft

Microsoft's CM suite solution includes Office 365 and SharePoint. The products offer robust functionality, integrate seamlessly with other Microsoft products, and are scalable for the enterprise. They also provide end-users with a centralized location in which to access content. Office 365 and SharePoint provide functionality for document management, collaboration, and automated workflows. Users can access content from other applications, create custom forms, work on files at the same time, and securely share content externally. Microsoft provides advanced enterprise-grade security, compliance, and governance through the Microsoft Security and Compliance Center. End-users can use Microsoft to conduct a GDPR readiness assessment.

Microsoft is well known for its frequent updates, and most users are familiar with the process. Upcoming plans for content services include predictive indexing and queries with AI, file move function to allow users to move content between SharePoint and OneDrive, and continued automation for approvals and publishing.

Nuxeo

Nuxeo is one of the newest vendors in the market but has established a strong market presence. Nuxeo is cloud-native, but it can also be deployed on-premise and in hybrid environments. One of the primary problems Nuxeo seeks to address is the issue of siloed content and data because most companies have multiple content repositories. Nuxeo addresses this problem by providing a platform that enables end-users to access and work with content from all different repositories through plugins to other CM solutions and to business applications such as Salesforce. The content from other repositories can be accessed in Nuxeo, enabling end-users to work with this content even as it stays where it is. This functionality also prevents the need for costly and time-consuming data migrations.

Nuxeo is committed to client-centric service and advances in usability. It plans to continue making improvements to the customer experience with additional usability features, AI enhancements to increase actionable insights, and the use of microservice architecture to design applications as suites of independently deployable services. Additionally, Nuxeo continues to develop partnerships to support platform offerings. Its growth strategy is aggressive, and it plans to be a global CM standard in the next three to five years.

OpenText

OpenText provides CM and BPM to a broad market of customers, many of whom are large enterprise companies. The OpenText Content Suite features robust out-of-the-box functionality and features a modern and intuitive UI in a drag-and-drop format. Business processes offered include accounts payable, human resources, asset management, and records management. The functionality of the product extends to information management, business networks, analytics, customer communications management, and eDiscovery. The solution also offers strong capabilities for governance, regulatory compliance, and security. Additionally, OpenText Magellan provides AI-based predictive analytics that combines open source machine learning with advanced analytic capabilities and business intelligence.

OpenText continues to expand beyond the CM market, offering solutions for many business requirements while adding investments in analytics, internet of things (IoT), case management, and cloud services. OpenText plans to grow via the acquisition of businesses that support digital transformation and process automation. The organization is planning to enhance content analytics for data blending and develop admin tools to make configurations easier. Deep learning image analysis, forensic threat detection, high volume data management, and hybrid offerings are all a part of the OpenText strategy. OpenText remains a leader in the CM market.

 ## ROI Case Study: OpenText

Energy Company • ROI: 378%

The company, an oil and natural gas developer and producer selected Documentum as-a-service (DaaS) with OpenText Cloud Managed Services to provide comprehensive enterprise content management (CM), significantly improving user productivity, redeploying some IT staff, and avoiding additional IT hires. With 300 users creating, filing, and searching for content, the lack of a unified CM platform impeded productivity. OpenText's Cloud Managed Services delivers the skills, support, and expertise necessary for an efficient DaaS, driving value to the company.

CUMULATIVE NET BENEFIT

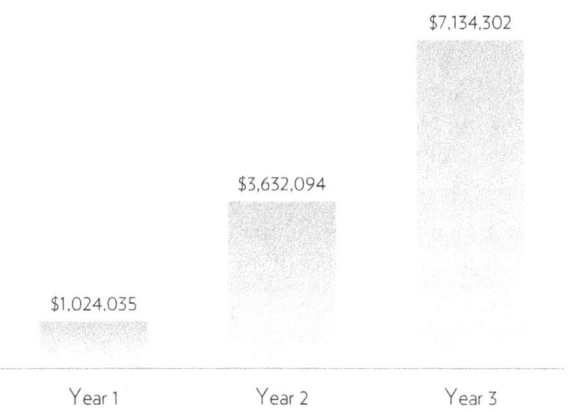

		$7,134,302
	$3,632,094	
$1,024,035		
Year 1	Year 2	Year 3

THE COMPANY
The company is based in the southeastern Pacific region,

operating oil fields and conducting oil and natural gas exploration. It holds an extensive portfolio and the company has a clear strategy to drive future growth with a wide range of exploration and appraisal activities.

KEY BENEFIT AREAS

Direct benefits revolved around the use of Managed Services from OpenText, off-loading a bulk of the IT work to the vendor. For this company's project, Cloud Managed Services eliminated costs for hardware and internal personnel resources. Indirect benefits of increased productivity were generated from the more traditional CM activities like a significant reduction of manual tasks for document management. Overall key benefits included:

- **Redeployed staff.** Previously, some staff's sole purpose was manually retrieving and verifying documents. OpenText's platform delivered easy access to documents and automatically provided verification, so those staff members were able to be redistributed to other areas of the business.
- **Avoided additional hires.** The OpenText Managed Services meant that the company could avoid hiring several administrators, since system and database administration activities are managed by Managed Services.
- **Increased productivity.** Productivity gains of 33 percent are attributable to the simplification of document retrieval and the ability to find the correct information, quickly.
- **Avoided hardware costs of on-premise deployment.** Once the company made the decision to purchase the DaaS subscription in the OpenText cloud, they were able to eliminate the expected costs of hardware and maintenance needed to support an on-premise solution.

BEST PRACTICES

OpenText Managed Services Cloud offerings eliminate

the end-user burden of managing applications and infrastructure needed for a comprehensive CM solution. Organizations with multiple locations and user groups require an efficient document and records management system that enables users to complete core functions more effectively, whether creating reports, diagrams, or other content.

· · · · ·

SER Group

The SER Group offers CM and BPM on a single platform to customers around the world. The SER Group is headquartered in Germany, and although most of its customers are in Europe, it is expanding its North American presence through its extensive partner network. Moreover, SER has its own sales representatives in Eastern Europe, Russia, and APAC. The vendor offers robust and scalable out-of-the-box functionality that allows for fast implementation, easy customization, and quick onboarding. The solution provides integrations to legacy systems so that users can access content from existing on-premise and cloud systems.

The vendor offers Doxis4, service-oriented architecture (SOA) with a substantial number of on-premise users. Doxis4 also runs in the cloud or on a hybrid platform. It provides CM and BPM on a unified platform, all of which have been developed organically by the vendor. Customers can purchase the solution as a full stack or select modular parts of it. The metadata-focused platform allows for content analytics. The solution also provides out-of-the-box case management tools with ready-to-use business solution templates for wholesale and retail, engineering companies, financial services institutions, insurance companies, hospitals, life sciences companies, M&A data rooms, federal authorities, and pension funds. Specific business processes include HR records management, accounts payable, quality management, claims management and mitigation, call center and ticket system solutions, contract management, and patient records management. End-users can invite others to access their document repository with limited permissions. The SER Group has many customers in regulated industries and provides strong governance capabilities, particularly for GDPR, FDA compliance, and government solutions.

SER Group has a limited North American presence, but plans to change this over the next few years. The company's value proposition is CM and BPM on the same platform so that it is scalable, flexible, and evolutionary. Its philosophy of "fulfilling customer needs" will play well as SER enters the North American market.

SpringCM

SpringCM offers a cloud-native platform with strong out-of-the-box functionality for document management, business process automation, and governance that allows for low implementation time. SpringCM offers strong capabilities for unstructured content management and has a strong presence as a solution for technology companies and industries with significant regulatory requirements, such as government and financial. The vendor offers business process automation for contract lifecycle management, client or employee onboarding, case management, regulatory compliance, and processing.

SpringCM has differentiated itself in the market with rapid deployments, simple integrations with API's and connectors, and deep functionality. Its product roadmap for 2018 includes a redesigned dashboard with additional navigation tools and task management, leveraging AI for analytical reporting, and records management. SpringCM plans additional enhancements to the user experience, in addition to its substantial functionality.

Analytics Value Matrix

Dot indicates current position. Arrow indicates trend in 2019 relative to others in the market.

CHAPTER 10
ANALYTICS

V endors on the Matrix are positioned based on the relative usability and functionality of their products compared to competitors and the overall market (Nucleus Research s142— *Understanding the Value Matrix*, September 2018). Advances in computing power and statistical techniques enable rapid development in the analytics space as companies are beginning to embrace data-driven decision making en masse. Nucleus sees three main trends driving investment in the Analytics market this year.

Embedded analytics. Vendors are increasingly moving to embed their analytics technology in third-party purpose-built applications as a way to smoothly integrate analytics into the workflow without users needing to navigate to a separate application or user interface. By embedding in other business applications, vendors can optimize algorithms for their intended use and data types, incorporate pre-built templates and models, and securely connect with data, all of which decreases the time to value for customers and improves overall usability. All Leaders in this Matrix and vendors with Leader potential are demonstrating mature embedding strategies.

Move to the cloud. More and more often, customers are eschewing on-premise data storage and technology management in favor of the agility and scalability of the cloud. As cloud technology has gotten more reliable and secure, companies are becoming more comfortable turning over their data to cloud providers and eliminating on-premise hardware (and their associated costs). With cloud-based analytics, customers can leverage cloud providers' high-performance computing power to drive advanced technologies like deep learning and real-time data processing. Nucleus found in its market survey, carried out in the first half of 2018, that over 60 percent of business intelligence (BI) deployments are in the cloud, with that number expected to grow (Nucleus Research s116 — *First half 2018 market survey*, August 2018).

Artificial Intelligence (AI). AI is finally turning a corner from amorphous marketing hype to an increasingly well-defined technology that delivers tangible business value. Vendors are taking advantage of vast data store to enable task automation like automated data preparation and cleaning. Additionally, predictive and prescriptive capabilities are becoming necessities for leading vendors as companies demand explanations for computer-generated insights and look to forward-facing analytics to prevent problems before they occur instead of traditional backwards-facing reporting. Looking forward, conversational AI will allow users to interact naturally with data through natural language processing (NLP) technology, and proactive signals-based intelligence will continue to gain value as a differentiator.

· · · · ·

Birst

Birst offers an enterprise-grade, cloud-based BI and analytics platform that delivers end-to-end capabilities to help organizations analyze complex business processes. It can be deployed on both private and public clouds and supports a variety of data sources, from ERP systems and Big Data platforms to popular cloud applications, spreadsheets and files. Built with patented automation and machine learning technologies, Birst's pioneering Networked BI approach connects centralized and decentralized teams and applications via a trusted network of analytics that share common business rules and definitions, enabling easier data blending and eliminating inefficient siloed structures. Birst is best known for its flexibility, with customers reporting that it is highly performant at scale on all types of data.

BOARD International

BOARD International is a software vendor based out of Boston, Massachusetts and Chiasso, Switzerland that offers a single product for corporate performance management, business intelligence, and predictive analytics. The solution can be deployed in a full-cloud or hybrid-cloud environment as well as on-premise. The BOARD platform enables end-to-end BI functionality including simulations, data exploration and insight discovery, data entry and management, and visualization.

Customers describe BOARD's product as highly usable and great for basic data reporting and modeling. Additionally, customers who want to simplify their technology portfolios may opt for BOARD as its solution supports both BI and CPM without needing to stack additional products. That said, the product lacks the functionality of some more advanced offerings, and customers have reported that online resources and documentation are lacking. It is positioned in the Facilitators' quadrant due to the product's high usability but relatively basic functionality for BI.

Domo

Domo for Business is a cloud-based analytics, visualization, and BI platform with intuitive dashboards that facilitate usability and do not require data scientists or IT for daily use. Users reported that its ease of use and rapid deployments are important factors in their choice of Domo. Domo offers easy data access and self-service analytics to support organizational agility and facilitate a cultural shift towards being a more data-driven company.

Its primary differentiators being collaboration with a host of collaborative tools built into the product to allow data sharing and explanation to make analysis explainable. Domo is gaining traction in the market and with continued development of more advanced capabilities such as additional machine learning and automation features, we expect it to continue to improve its positioning.

Dundas

Dundas BI is a is a browser-based business intelligence and data visualization application that offers interactive custom dashboards, ad-hoc queries, visual reporting tools, and data analytics. It is consistently described by customers as being easy to implement and use. Dundas BI can be deployed as the central data platform for a business or can be embedded into an existing application. Dundas BI security supports multi-tenant scalability SaaS deployments.

As a smaller standalone vendor in an increasingly crowded market, we believe Dundas will be challenged to remain viable moving forward without substantial investment in industry-standard capabilities such as AI and advanced, interactive data visualizations.

Information Builders

Information Builders WebFOCUS is a full-suite, interactive BI and reporting platform that converts business data into actionable insights, using machine learning (ML) with AI capabilities for predictive analytics and Natural Language Queries (NLQ) in addition to traditional BI. WebFocus is a scalable solution with seamless integrations with multiple platforms such as Office 365. Nucleus found when interviewing end users that WebFOCUS requires reasonably sophisticated IT support and is best-suited to larger enterprise deployments.

WebFOCUS is a highly functional platform with high-performance capabilities, thereby making it a good choice for enterprises with large volumes of data. Customers mentioned a distinct learning curve to become self-sufficient on the platform, which along with the necessary IT support, impacts the usability score. Information Builders is placed in the Leader quadrant of this Matrix for the enterprise-class functionality it delivers. With additional investment in usability, it is poised to improve its positioning in future Matrices.

GoodData

GoodData is an end-to-end cloud-based predictive analytics platform that delivers insights in context at the point of work. The platform is secure and scalable with multi-tenant distribution and is also offered as a  fully managed service. GoodData continues its focus on enhancing usability and reducing the need for extensive IT resources. Users can build insightful reports and visualizations that support better business decision and benefit from GoodData's expertise in consolidating, transforming, and distributing actionable data.

The GoodData platform is proven to deliver value, as GoodData deployments have been recognized in our annual ROI Awards for the exceptional returns that customers received. With this and its recent investments in automation and other usability improvements, GoodData keeps pace with the market in terms of functionality and advances in usability. We believe

GoodData delivers significant value and fully expect it to continue to succeed in the market.

ROI Case Study: GoodData

EmeraldCube Solutions • ROI: 1273%

EmeraldCube deployed GoodData to provide business intelligence (BI) services to its customers through a white-labeled solution, EmeraldVision. The company chose GoodData over a self-built solution, saving an expensive investment in capital and personnel, as well as two years' worth of work. Within three months, EmeraldCube was able to offer its customers a cost-effective BI solution that readily delivers visibility into business operations and automates manual processes. As a result, EmeraldCube has increased its revenue, improved customer satisfaction, and avoided infrastructure costs and hiring additional staff necessary to support a self-built system.

CUMULATIVE NET BENEFIT

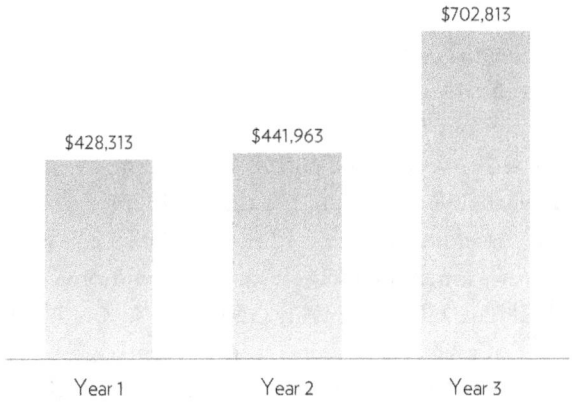

THE COMPANY
EmeraldCube Solutions provides services and solutions for Oracle's JD Edwards enterprise resource planning (ERP)

application. The company was founded in 2013 and is headquartered in Alpharetta, GA. Within its managed services and project services offerings, the company offers support for the JD Edwards application, development support, Oracle BI support, and support for upgrades and migrations. EmeraldCube also provides proprietary third-party solutions for JD Edwards, including solutions for system monitoring, encryption, and fraud detection. The company's most robust proprietary solutions include EmeraldCloud, a managed cloud for JD Edwards through Amazon Web Services (AWS), and EmeraldVision, a BI solution for JD Edwards through GoodData.

KEY BENEFIT AREAS

Deploying GoodData through EmeraldVision helped EmeraldCube increase revenue, avoid the significant costs of a self-built solution, increase customer satisfaction, and improve its customer to employee ratio. Key benefits of the project included:

- **Increased revenue.** All EmeraldVision customers are net new business for the company. Because the implementation cost and total cost of ownership (TCO) of GoodData is low, there is a fast time to value for customers. For the company, there is a high margin on the recurring revenue that comes in from its customers' subscriptions to EmeraldVision.
- **Avoided costs of a self-build.** By deciding to buy and white label GoodData rather than building a solution itself, EmeraldCube avoided the significant costs and time investment of a self-build. A self-built project would have taken two years, and the costs would have included a significant upfront investment as well as the costs for 4 developers and 1 data architect to build the solution and support it on an ongoing basis.
- **Increased customer satisfaction.** There has been a zero percent churn rate of all EmeraldVision customers. End users get insights delivered in an average of one

minute, while competing solutions can take as long as 10 minutes. Customers benefit from the low implementation time for EmeraldVision — deployment can take anywhere from a few days to a month — and the fact that it is an end-to-end solution that requires minimal support. Customer success stories include one customer who has saved significant amounts of time by automating a manual sales lead follow-up process, and one manufacturing customer who has increased company performance by producing on-demand branch reports and regularly assessing performance relative to quarterly goals.

- **Improved customer to employee ratio.** Because EmeraldVision is SaaS-based, it typically only requires the support of one employee per deployment. As a result, the company is better able to offer fast, individualized support to its customers. This has enabled the company's employees to support more customers and to focus their time on other high-value aspects of their work.

BEST PRACTICES

Deploying EmeraldVision as a white-labeled version of GoodData allowed EmeraldCube to revolutionize its go-to-market BI strategy. With GoodData, EmeraldCube was able to deploy BI to its customers quickly and cost-effectively through EmeraldVision. In turn, its customers benefit from a low TCO and the ability to access insights quickly to improve company performance.

• • • • •

IBM

IBM is a global enterprise software and hardware provider that was founded in 1911 and is based out of Armonk, New York. IBM offers several products in the data analytics and BI space, so this report will focus IBM Cognos, its traditional enterprise-scale BI platform.

Cognos tends to be used in composite with other analytics tools; however, its maturity, reliability, and relatively large user ecosystem makes it a reliable choice for customers seeking traditional large-scale BI.

We have been quick to level criticism at IBM before for being slow to reflect market trends and for confusing messaging around its products and capabilities. We have found Cognos to be highly functional and deliver legitimate value to users; however, some customers we interviewed did not understand which IBM product to use, impacting IBM's overall footprint in the space. With these most recent product updates along with a branding correction that clearly positions Cognos as the enterprise BI tool to use, IBM is demonstrating its recognition of market demand for integrated intelligence within BI tools and vertical-specific solutions, and is creating a legitimate roadmap to achieving positive ROI. Cognos delivers full-service BI capabilities at enterprise scale and, as its usability continues to improve with future investment, we expects IBM to advance in the Matrix.

Information Builders

Information Builders WebFOCUS is a full-suite, interactive BI and reporting platform that converts business data into actionable insights, using machine learning (ML) with AI capabilities for predictive analytics and Natural Language Queries (NLQ) in addition to traditional BI. WebFocus is a scalable solution with seamless integrations with multiple platforms such as Office 365. Nucleus found when interviewing end users that WebFOCUS requires reasonably sophisticated IT support and is best-suited to larger enterprise deployments.

WebFOCUS is a highly functional platform with high-performance capabilities, thereby making it a good choice for enterprises with large volumes of data. Customers mentioned a distinct learning curve to become self-sufficient on the platform, which along with the necessary IT support, impacts the usability score. Information Builders is placed in the Leader quadrant of this Matrix for the enterprise-class functionality it delivers.

Jinfonet Software

Jinfonet's JReport is an embedded analytics solution with customizable reporting, analytics, and dashboards built for enterprise-class scalability and performance. JReport's deployment is available in the cloud, as-a-service, and on-premise, and can be customized to any business application within the context of existing software. Its Self-Service Analytics allows users to build ad-hoc reporting and dashboards from any location and analyze data with or without IT assistance. In the past year, JReport has focused on enhancing usability and its sophisticated feature sets in response to user feedback.

Jinfonet's product is highly functional with feature highlights including its scalable and performant architecture, clustering capabilities, multi-tenancy, and advanced reporting. As an embedded solution, overall usability is limited compared to some mature standalone products; however, with continued success in the market and additional reinvestment in improving its offering, we believe Jinfonet shows the potential to contend for a position as a leader among embedded analytics vendors.

Logi Analytics

Logi Analytics is an embedded visual and predictive analytics platform focused on application teams, specifically product managers and developers. Logi enables companies to match their data analytics reporting and visualizations to their existing user interfaces (UIs). Logi connects to all relational databases, BigData, and company files and delivers security protections through multiple changes including changing its focus to exclusively emedding its analytics technology whether in the commercial software market (OEMs, ISVs) or for custom enterprise applications. Additionally, in late 2017 Logi was acquired by Marlin Equity Partners. Logi's analytics development platform delivers extensive functionality in all areas of predictive analytics, security, connectivity, and architecture, all designed for application teams.

Logi focuses on giving customers control of their analytics by integrating with their data, customizing with their branding, and deploying on their hardware. It scales well without demanding exorbitant compute power, a key value point for embedding in third-party software, and can be easily governed. Recent investments have been in the deployment process and machine learning capabilities to empower non-data scientist users. Logi is positioned in the Experts' quadrant of this Matrix because it delivers highly customizable and advanced analytics capabilities, but it does require a development team to configure and deploy.

Looker

Looker is a big data vendor that that uses SQL to access and manipulate data, allowing users to explore, integrate, and visualize data in real-time using a browser. It is primary deployed in the cloud but can be deployed on-premise. Looker connects to any relational database and automatically generates a data model from the information that can be customized to a company's specific needs. Looker offers an embedded option as well as a standalone application and uses LookML language to build SQL queries against a specific database. Customers have described Looker as being a data scientist-driven application, but it has expanded its self-service exploration capabilities and is usable for non-technical users as well with training.

This past year Looker has focused its development on functionality, which may limit its deployments to those businesses with more sophisticated IT resources or BI teams. As the company grows and is able to reinvest profits to make the product easier-to-use, we expect Looker to become more prominent in the market.

Microsoft

Microsoft PowerBI is a suite of business analytics tools that offer self-service data management, analysis, querying, visualization, and dashboarding tools to business users of all kinds. Microsoft is a global software provider based in Seattle, Washington and its products are designed to be highly usable and scalable from a single user to a full enterprise. The products connect data sources from the Web, databases, and independent software vendors to allow businesses to combine and understand their data from every channel. PowerBI comes as a standalone free desktop application but can also be hosted on the cloud for sharing, schedule refresh, and friendly data consumption experiences, as well as embedded in other tools. The products include Power BI Desktop, Power BI Pro, Power BI Premium, Power BI Mobile, Power BI Embedded, and Power BI Report Server.

Microsoft regularly updates the products each month with additional functionality and continual UI improvements to drive increased usability. These updates are influenced directly by users' feedback and requests. The interface and controls are similar to those of Microsoft Excel, with which many legacy BI users have experience.

Power BI is a powerful, usable BI option that is integrated with the largest number of third-party data sources on the market. Regular updates to product functionality, including significant investment in edge technologies like AI, computer vision, and natural language processing, and a highly usable interface justify its positioning in the Leaders' quadrant of the Matrix.

MicroStrategy

MicroStrategy offers a mature BI solution with end-to-end capabilities for data ingestion and management through to visualization and dashboarding. MicroStrategy is a well-established BI provider with strong capabilities in reporting, dashboarding, and data analysis. It is traditionally deployed on-premise although it has expanded to offer cloud hosting as well. The product is built on a complex infrastructure that can pose usability challenges to customers, particularly those without technical training, as the product is light on self-service features.

The company has added 196 new employees over the last year and is continuing to invest in developing its technology. With successful releases of Version 11, MicroStrategy is poised to advance in the next Matrix. In particular we note the significant functional improvements the company made to the product in the past year.

Oracle

Oracle is a leading provider of Cloud solutions founded in 1977 and based out of Redwood Shores, California. Oracle provides Cloud solutions for all major business functions like CRM, ERP, BI, and HCM. Oracle helps customers develop roadmaps, migrate to the cloud, and take advantage of emerging technologies from any point: new cloud deployments, on-premise environments, and hybrid implementations. Oracle's approach makes it easy for companies to get started in the cloud and even easier to expand as business grows. Its analytics and BI product, called Oracle Analytics Cloud can perform all major BI tasks such as visualization, self-service reporting, data preparation, scenario and what -if analysis, and advanced analytics. Keeping pace with market progress, Oracle also offers two cloud models, Oracle Analytics Cloud and Oracle Business Intelligence Cloud Service.

With a long history in the software business, first as a database company, Oracle has an established customer base and the industry experience to deliver great value. However, Oracle on-premise products are unwieldy to integrate with third-party products, and customers have said that it only makes sense to choose Oracle for BI and analytics for companies already using the Oracle E-Business Suite due to the high cost and difficult integration. Although the product itself is powerful for traditional BI tasks, we position Oracle in the Core Provider quadrant of the Matrix due to the difficulty of integrating with other tools and the fact that customers feel that it is a feasible choice only if the rest of the business runs on Oracle too.

Qlik Technologies

Qlik is a cloud-based self-service analytics and business intelligence solutions provider. Many customers deploy Qlik as an end-to-end BI solution, handling data ingestion and preparation as well as analysis, visualization, and reporting. Its main products include Qlik Sense and QlikView, in addition to a portfolio of add-on products for specialized functionality. A main value differentiator for Qlik is its Associative Engine, an AI-powered tool that indexes relationships between data.

This year, In September, Qlik announced two new Big Data products: an updated release of the Podium Data product as well as the first-time release of the Qlik Associative Big Data Index. The Podium update contains two key new features, a catalogue module and an intelligent rules engine. New metrics within the data catalogue allow users to "shop" for data as well as identify the most useful insights. The Intelligent Rules Engine analyzes data during ingestion and stores or acts on it based on the analysis. The Associative Big Data Index delivers associative insights on top of large-scale big data sources. Docker containers allow users to analyze big data at its source and remain platform agnostic, and computing upgrades optimize the system to perform on massive amounts of data without a drop-off in performance.

As a standalone analytics and BI provider, Qlik is challenged by the increasing market preference for embedded BI solutions that are built on top of existing business platforms. That said, by entering strategic partnerships with independent software vendors (ISVs) and continuing to invest in delivering increased usability and functionality, Qlik remains poised to be a major player in the BI market in the year to come.

Salesforce

Salesforce is a global cloud software company based in San Francisco. It offers business applications for sales, marketing, and service with a specialty in customer relationship management (CRM). Einstein is Salesforce's AI engine built on top of its business platform. Einstein Analytics is Salesforce's standalone analytics product which enables full-service analytics from basic reporting and dashboards through advanced and AI -augmented data analysis. Einstein Analytics is natively integrated with customer data and activities in real-time, so it delivers self-service BI functionality within the Salesforce workflow to users. The Einstein engine is also available as a standalone product. Developers can leverage the AI built by Salesforce and create custom applications on the Einstein platform.

Salesforce differentiates its analytics approach in two main ways—first, AI is integrated with traditional analytics to allow for task automation and a wide breadth of functionality, and second, its analytics technology is embedded within the CRM clouds to support users in making contextualized, data-driven decisions within the workflow.

Salesforce continues to demonstrate significant investment in its analytics platform with regular functional upgrades, although it should take care that as more capabilities are added, product usability doesn't suffer. With the addition of Einstein Voice, the product usability is significantly increased as users can interact with Salesforce natively in a non-technical avenue through speech. With high levels of product functionality, including predictive and prescriptive analytics, a user-friendly, code-free interface, and readily available training materials via Trailhead, Salesforce is placed comfortably in the Leaders' quadrant of this Matrix.

SAP

SAP was founded in 1972 in Walldorf, Germany. It is historically a major player in enterprise software, offering solutions for the full range of business functions such as ERP, CRM, and BI. It maintains an expansive legacy user base with large-scale on-premise deployments but also offers fully cloud-based and hybrid cloud solutions. Its products for BI and analytics include SAP Analytics Cloud, SAP Predictive Analytics, SAP Analytics Hub, and SAP Leonardo Machine Learning.

SAP products are diverse and mature, ensuring deep functionality for a wide range of use cases and an experienced user community. Usability has always been a weak point with users consistently mentioning a steep learning curve and the need for extensive internal IT support. Data governance and integration issues also can come into play with the wide range of separate modules that are offered. SAP is best-suited for enterprise-scale deployments where the full business can be outfitted with SAP products that integrate at the backend, but it may struggle to remain competitive against vendors offering more easy-to-use products and where customers are choosing point solutions or embedded analytics.

SAS

The SAS suite was developed by the SAS Institute, a global software company based out of Cary, North Carolina that was incorporated in 1976. It offers a suite of data management and analytics solutions that can mine, alter, manage, retrieve, and analyze data. Its primary focus area is customer intelligence; however, other products handle a variety of specialized use cases including: fraud detection, risk assessment, IT management, and KPI monitoring. Additionally, it offers industry-specific solutions for government, retail, telecommunications, aerospace, marketing automation, and high-performance computing.

Stemming from its many decades in the market, SAS retains a significant legacy user base. It is widely recognized as the leading platform for advanced statistical analysis due to its comprehensive functionality. Additionally, SAS is used in academia for research and education, which ensures that there is a steady flow of trained users on the platform. Due to its complex product list and the steep learning curve needed to become competent with the software, SAS is better suited for large enterprise customers with the resources and knowledge base to maximize the value of the investment.

Sisense

Sisense is New York-based company that delivers an end-to-end BI platform that specializes in collaborative dashboards and data visualizations. It is sold as a single tool solution that is equipped to ingest, prepare, analyze, and report on data of all formats and amounts. In addition to being a standalone product, Sisense can be embedded or white-labeled in third-party products, expanding its footprint.

In addition to investing in new product features and its growing sales force to drive the business' expansion, Sisense is also aggressively developing cloud infrastructure and additional AI-powered capabilities. Sisense marries advanced functionality with a highly usable interface, justifying its position in the Leaders' quadrant.

Tableau

Tableau is a BI provider based out of Seattle, Washington. It specializes in data visualization and dashboarding. The products are integrable with third party tools so that Tableau visualizations are easily embeddable and are known to be easy-to-use, which allows for quick user adoption. Tableau software is available as a desktop application and as a Web application hosted on the cloud. Its products include Tableau Desktop, Tableau Prep, Tableau Server, Tableau Public, and Tableau Online, in addition to its embedded offerings.

Tableau has one of the most mature BI offerings available on the market with one of the largest customer footprints. Its products are highly usable and easily scale from the individual user to the complete enterprise. Functionally, its products focus more on visualization and dashboards than heavy-duty analytics, data manipulation, and next-gen technology like AI; however, it still offers sufficiently advanced functionality to justify its position in the Leaders' quadrant of this Matrix, particularly with recent announcements around upcoming NLP and data modeling capabilities.

TIBCO

JasperSoft by TIBCO offers an interactive, embedded dashboard and reporting solution with broad flexibility for data visualizations and reporting. JasperSoft delivers actionable insights with self-service for users to identify data they need inside their exiting application. It is scalable and can connect directly with big data such as with native reporting and analytics in real-time.

Yellowfin

7 is an Australian BI provider that offers an analytics platform with three main focus areas: dashboarding, visualization, and storyboarding. The platform has additional tools to support data integration and preparation as well as data alerts, a mobile application, and collaborative tools. Customers describe the platform as being very user-friendly with highly accessible on-line resources and documentation for self-study. The product suffers when

compared to more comprehensive end-to-end offerings that have more powerful functionality for data extract-transfer-load (ETL) and continued development is needed to enable support for more reliable analysis of unstructured data.

This year, Yellowfin announced Yellowfin 7.4.6 which contained a number of product improvements. The Assisted Insights feature has been upgraded to include comparisons and explanations for each insight generated. Data visualization is improved with improvements to Time Series manipulation. Responding to some customer issues, the update includes improvements for data preparation and transformation as well as backend performance upgrades to enable more competitive data management functionality.

Altogether, Yellowfin offers a reliable single platform BI tool for companies that don't have extensive data integration or management needs. The product focuses more on visualization and display than on hardcore functionality like predictive and prescriptive analytics, and this focus on usability justifies its positioning in this Matrix.

Zoomdata

Zoomdata offers an innovative BI solution. The company's high-performance BI engine and visual analytics allow users to discover new opportunities and solve problems that are too big or too complex to solve using conventional BI tools. Zoomdata's interactive dashboards, native modern data connectors, scalable microservices architecture, and innovations such as Zoomdata Data Sharpening make it a natural front-end for big data, live streaming data, and multi-source analysis. Launched in 2014, Zoomdata holds multiple patents related to streaming data delivery and interactivity.

Zoomdata differentiates itself with its data handling and streaming capabilities, making it a natural choice for real-time or low-latency analytics. It lacks the advanced AI-powered capabilities of some more comprehensive products, but Zoomdata can easily handle the demands of traditional BI without sacrificing performance or usability and is positioned as a Facilitator in this Matrix.

Corporate Performance Management Value Matrix

Dot indicates current position. Arrow indicates trend in 2019 relative to others in the market.

CHAPTER 11
CORPORATE PERFORMANCE MANAGEMENT

Two main trends persist in the corporate performance management (CPM) market: cloud uptake remains consistent but slow, and many cloud vendors are divided about breadth of functionality. While the financial planning and accounting teams remain separated in their software choices, CPM products are branching out in functionality to cover many use cases. However, user adoption of a single company-wide CPM solution remains low.

Cloud CPM solutions have been generally seen as replacements for either Excel or legacy on-premise products. Recently, customers are finding a more crowded marketplace and greater competition among cloud vendors as functionality expands to more diverse use cases. Customers now have a range of deployment options for their CPM solution: a modularized on-premise product, multiple cloud products, or a single diverse cloud application.

There is still confusion around how to build a best-of-breed stack. Many vendors start with specialized financial services and develop functionality from there. However, we discovered that many accounting and financial planning teams do entirely separate software evaluations and deployments. The often-successful land-and-expand paradigm of departmental deployments does not yet show the promise of high ROI for CPM.

Relative to other software areas, many customers still deploy on-premise solutions, despite sometimes considering cloud vendors. Most finance teams heavily prioritize consistency and familiarity over innovation. As long as a chosen software works, CPM customers will usually stay with it. ERP systems are very difficult to change, and CPM applications by extension have their own integration challenges that can complicate transitions. Finally, because financial applications need a near-constant uptime, customers who want to change vendors have to run products in parallel, which can be laborious and costly.

The trend of slow cloud uptake shows a lack of clear understanding of the cloud's advantages and the stickiness of finance culture. Indeed, many customers simply look to their ERP provider to adequately fulfill their CPM requirements. We expect that customers will soon be better equipped to evaluate their needs and vendor offerings, resulting in easier and less expensive transitions to the cloud.

In this Value Matrix, we analyzed ease of use for end users and ease of deployment. We also evaluated a broad range of functionalities including planning, budgeting, and forecasting (PBF); modeling and what-if scenario generation; reporting; financial close and consolidation (FCC); account reconciliation; risk and audit management; disclosure management; and data integration with ERP, HCM, or other systems.

· · · · ·

Adaptive Insights

Adaptive Insights was founded in 2003 and is headquartered in Palo Alto, California. Adaptive Insights has been a long-time leader in the CPM market, and with its cloud-only product, commands a strong customer base of first-generation cloud CPM deployments. The product is browser-based and integrates with Excel through Adaptive Insights' OfficeConnect. In contrast to some other Office integrations, OfficeConnect focuses on using pre-built templates to minimize the amount of building that end users need to do when pulling together data. Adaptive uses its own in-house integration platform to connect with ERP, CRM, HCM, and payroll. It also offers open APIs (application programming interfaces) for customers to connect their systems through existing integration platforms like Dell Boomi or other common third-party options.

Adaptive Insights has its strongest presence in the North American mid and downmarket, with a sizable chunk of US enterprise and global customers. It offers both accounting functionality and core planning tasks, although we found that deployments are often initially focused on either the accounting or FP&A teams.

In 2017, Adaptive Insights released several product updates, including self-service analytics, a process guide for infrequent users, advanced formula expressions, flexible time modeling, and more constant currency reporting for consolidation.

Adaptive Insights has strong PBF, modeling, and reporting. Customers found the product to be intuitive, especially for non-technical users, and that the Excel Interface for Planning gives flexibility for different users and needs. Most customers found deployments to be easy, although we received some mixed feedback.

Overall usability was high across planning, collaboration, and what-if scenario generation. Customers reported that support was also generally positive. Although price was not a factor in Matrix placement, many customers found Adaptive's pricing to be a strong positive factor in their decision to purchase.

In 2018, Adaptive Insights is continuing its focus on building out greater functionality for advanced modeling and financial tasks, while maintaining good usability.

However, as more customers move off legacy on-premise products in favor of the cloud, and as some customers move to their second generation of cloud CPM products, Adaptive Insights will find greater competition from new vendors competing for its cloud market share.

 ## ROI Case Study: Adaptive Insights

Creation Technologies • ROI: 144%

Creation Technologies deployed Adaptive Insights to improve its financial reporting and forecasting. Moving from a legacy on-premise system has enabled Creation to increase the accuracy and consistency of its data, increasing productivity across all levels and tasks of the finance department. Additionally, Creation has been able to reduce its costs related to third-party integration and lower the cost of administrative support. The self-deployment model that Creation elected to implement also created a greater understanding of the technology and how to use the solution to meet the company's needs.

THE COMPANY

Creation Technologies provides original equipment Manufacturers (OEMs) around the world with integrated (Electronics Manufacturing Services) EMS solutions that help customers accelerate time-to-market, reduce operating costs and grow revenue. The company is headquartered in Burnaby, British Columbia, Canada. Creation provides full service solutions across North America and worldwide, and has 11 manufacturing operations in Canada, the USA, China and Mexico.

CUMULATIVE NET BENEFIT

$686,412

$304,828

$59,244

Year 1   Year 2   Year 3

KEY BENEFIT AREAS

Deploying Adaptive Insights to improve its financial reporting and forecasting operations enabled Creation Technologies to reduce third party support, increase productivity, and increase accuracy. There were various benefits of the project:

• The company was able to reduce third party support by directly educating the administrators about the set-up and technology so that it no longer needed the outside help necessary for its old system, Hyperion.
• The company's board regularly asked for updates, a change in reporting format or organizational structure

and different "what if" cases. With Adaptive, this task now requires considerably less effort, as the system does most of the heavy lifting.

- Creation is now also able to do vendor integrations with its own staff in its own time.
- Finally, the company's financial reporting team was able to increase its productivity because the data being used is reliably accurate, meaning reports do not have to be redone, or numbers verified.

BEST PRACTICES

Creation Technologies did a substantial amount of pre-work to map out what it wanted, including designing a new organizational structure, separate models and user-friendly interfaces for input. When the implementation process started, however, this documentation stopped, as more time and resources were put into the implementation. Because of this, the company's model is not documented anywhere. Now that the go live has happened, and employees are comfortable with the solution, the company is working out a way to document the model somewhere.

Creation's main advice to others was to keep everything as clean as possible. Initially, the company was not keen on deleting anything that was already done. To keep the system clean, the company has segregated the old data in a separate folder. Creation is also big proponent of the self-implementation, as it felt it gave the company greater control on understanding and maintaining the solution in the future.

· · · · ·

Anaplan

Anaplan was founded in 2006 and is headquartered in San Francisco, California. Anaplan finds itself in the Leader quadrant because of its powerful and scalable functionality. Anaplan is a unique vendor in the CPM space as it offers an open platform for individual customers to build out their needed functionality at scale.

The product is significantly different from the majority of Anaplan's competitors' offerings and takes a different approach to addressing CPM needs. The in-memory engine, called Hyperblock, is designed for the cloud and allows Anaplan to compete well in complex, enterprise-scale scenarios traditionally limited to the advanced functionality of legacy on-premise solutions. Anaplan typically competes in the upmarket where customers have the internal support and bandwidth to deploy and customize the solution. Anaplan's focus is on FP&A, planning, budgeting, forecasting, modeling, and reporting. Not many customers have implemented Anaplan for financial close and consolidation or disclosure management.

In January 2018, Anaplan announced a new data encryption security measure, called Bring Your Own Key. The product is designed to allow more self-service security functionality for customers using the product in the cloud. In Q3 of 2017, Anaplan released enhancements to its Application Lifecycle Management (ALM) capability around production imports and difference reporting. ALM helps customers move applications between development and production environments and improves security, governance, and efficiency.

The Anaplan platform is highly flexible, and customers consistently remarked that the product maintains speed with large data sets. Functionality for PBF, modeling, reporting, and data integration is quite robust. Anaplan releases major product updates quarterly and minor updates monthly, and customers noted that the vendor responds to user requests and incorporates suggestions into future releases. Customers cited the Anaplan App Hub as a positive point within the user community.

Anaplan is continuing to increase in its functional strength, especially for large-scale customers looking for a robust cloud solution. Given the focus on usability in the Anaplan roadmap, we see them keeping pace with the market in usability and improving in functionality

BlackLine

BlackLine was founded in 2001 and is headquartered in Woodland Hills, California. BlackLine dominates the market for advanced accounting functionality around close and consolidation, account reconciliation, and compliance and risk management. BlackLine is different from many other vendors in the Value Matrix in that it does not directly provide core PBF or other related functionality. As such, BlackLine is positioned in the Facilitator quadrant due to its specialization in one branch of performance management functionality and not on the basis of some simplicity in the product.

In November 2017, BlackLine announced platform enhancements including a new user interface and new dashboards, improved automation for transaction matching, and additional public APIs.

BlackLine is a market leader in financial close and consolidation and account reconciliation. Users commented on the robustness of the product and the high value they derived from reduction of manual accounting processes. Users also noted that the product was easy to use with an appealing interface. Customer support was generally good.

BlackLine demonstrates continued success among its customer base and is expanding into the market organically and through a partner ecosystem with more traditional CPM vendors. We have received mixed feedback from customers about their preferences for integrated financial and strategic functionality within a single CPM product. Such preferences will impact BlackLine's popularity in the midmarket where such dual-functionality products are most viable. In any case, BlackLine will likely improve in usability but will decrease in overall functionality (both FP&A and accounting together) relative to the market as more planning vendors build out accounting functionality within their products.

 ## ROI Case Study: BlackLine

Sirius XM • ROI: 95%

A long-time BlackLine customer, SiriusXM recently deployed BlackLine's automated journals capabilities to improve the efficiency of its accounting department and ease the review processes associated with the financial close. Leveraging the cross-product functionality of BlackLine, SiriusXM was able to automate 50 percent of its journal entries within two years and expects to automate another 20 percent within the current year.

THE COMPANY
SiriusXM Holdings is a leading audio entertainment

company that delivers a suite of subscription content through both satellite and online radio. Formed after the merger of XM Satellite Radio and Sirius Satellite Radio in 2008, SiriusXM has grown its subscriber base from 18.8 million in 2009 to over 32 million in 2017. Partnering with automakers and car dealerships, SiriusXM is available on over 70 percent of new vehicles entering the market.

CUMULATIVE NET BENEFIT

$440,224

$181,294

$64,344

Year 1   Year 2   Year 3

KEY BENEFIT AREAS

SiriusXM worked with BlackLine to automate many of its accounting processes, which significantly improved the efficiency of the finance department.

- With BlackLine, SiriusXM has automated approximately 50 percent of its journals. The accounting department expects to expand the percentage of journals that are automated moving forward. As transactions are captured and automatically imported into BlackLine, the system matches each transaction with the appropriate journal entry or other subledger transaction. BlackLine flags any exceptions for review. SiriusXM has achieved 99.9 percent accuracy in its matching automation process.
- SiriusXM has configured rules that enable low-risk accounts, journals, and tasks to be certified without

manual review. By automating multiple steps and systematically accomplishing tasks related to the financial close throughout the month, SiriusXM has reduced the time it takes to reconcile its accounts and produce reports for decision-makers and compliance.

- Leveraging multiple products, SiriusXM regularly loads data into BlackLine such as bank transactions, credit card payments, and general ledger and subledger data. The system automatically books transactions and matches them with their journal entries, reducing the amount of manual review required and the risk of errors from manual data entry and calculations.

BEST PRACTICES

Leveraging BlackLine's cloud deployment helped SiriusXM actualize its project goals in two ways: first, SiriusXM was able to leverage significant BlackLine functionality already deployed (e.g. Account Reconciliations, Matching, Tasks, and Journals); and second, SiriusXM's accounting department was able to complete the project without using significant IT resources, streamlining the overall project timeline and ensuring that the project was responsive to the accounting team's needs.

SiriusXM did not incur any ongoing maintenance costs from its automation project because employee time after going live was a value-add to the organization. Managers spent time ensuring that the solution produced results which were accurate and required no day-to-day upkeep that was not additive to the project.

· · · · ·

BOARD

BOARD International was founded in 1994 and has co-located headquarters in both Chiasso, Switzerland and Boston, USA. BOARD has historically had its strongest presence in the EMEA region, but with the addition of its Boston headquarters in 2016, it has been making a concerted effort to break into the North American market. BOARD's product differentiates itself from many other CPM products because of its single application approach and joint focus on business intelligence (BI) and CPM functionality within the same interface, and because the product is exactly the same on-premise and in the Microsoft Azure cloud.

In September of 2017, BOARD released its latest update to the financial close and consolidation module with a new user interface and several functional improvements. BOARD has a solid roadmap and is positioned to take advantage of new trends in advanced analytics and modeling, as well as greater end user selfservice.

BOARD customers consistently praised the product's usability and customer support. Because of BOARD's wide range of functionality for traditional BI, PBF, and reporting, customers tend to deploy across multiple departments rather than use a land-and-expand strategy.

In 2018, we expect to see continued expansion in the North American market and will be following closely how BOARD progresses as the legacy on-premise vendor market share further divides among cloud offerings. We see them keeping pace with the speed of the market in usability and functionality over the next year.

CCH Tagetik

CCH Tagetik, now a part of Wolters Kluwer, was founded in Italy 1986 and is headquartered in both Lucca, Italy and Stamford, Connecticut. CCH Tagetik has a strong EMEA presence and continues to expand its customer base in North America. Historically, the vendor was sought after for its functionality in end-to-end FCC.

Today, CCH Tagetik's Financial Performance Platform provides functionality for PBF, management reporting, predictive forecasting and modeling including what-if scenario playing, analytics and dashboarding, disclosure and regulatory reporting, and compliance and risk management. It offers many industry-specific solutions that include banking and financial services, insurance, manufacturing, retail, telecom, healthcare, and more. The vendor's customer base mostly consists of midsized and enterprise customers. CCH Tagetik offers identical solutions both onpremise and in the cloud. SaaS (software as a service) deployment is on AWS or it can be hosted by Microsoft Azure. On-premise, it runs on Microsoft SQL Server, Oracle, or SAP HANA databases. However, the majority of the vendor's new clients choose its cloud solution. On the accounting and revenue side, CCH Tagetik competes with Oracle and SAP. In the PBF space, CCH Tagetik competes with Host Analytics, Anaplan, Oracle, and IBM.

In April 2017, Wolters Kluwer Tax & Accounting completed its acquisition of Tagetik. In May 2017, CCH Tagetik announced the release of its IFRS solution portfolio, including its CCH Tagetik IFRS 9 and CCH Tagetik Revenue Accounting (IFRS 15), CCH Tagetik Lease Accounting (IFRS 16), and CCH Tagetik IFRS 17 solutions. These products include a contract data repository, compliant data models and calculations, and pre-configured workflows.

Customers note that CCH Tagetik offers a full range of functionality from a single platform. Customers noted the advanced regulatory and XBRL capabilities of the product. CCH Tagetik differentiates itself with its cash flow and operation planning modules, as well as a range of specialized, pre-packaged disclosure and compliance software including solutions for Solvency II including Pillars I, II, and III; Financial Instruments (IFRS 9); Revenue Accounting (ASC 606 & IFRS 15); Lease Accounting

(ASC 842 & IFRS 16); Insurance Contracts (IFRS 17); European Banking Authority (EBA) supervisory reporting; Basel III; Annual Reports; Budget Books & Board Books; and Integrated Reporting (IR). Customers also find the vendor's usability high because it has a drag-and-drop interface and provides a collaborative platform based on workflows. CCH Tagetik offers cumulative and non-disruptive update packets every quarter and major releases every one-and-a-half to two years.

In 2018, CCH Tagetik will continue to expand in the North American market. We also see CCH Tagetik focusing its efforts on compliance and regulatory reporting, advanced modeling and planning, integration and flexibility, data and process governance, and predictive analytics. Over the next year, the vendor will keep pace with the market in functionality and outpace the market in usability.

Centage

Centage was founded in 2002 and is headquartered in Natick, Massachusetts. Its CPM product, Budget Maestro, provides core planning, budgeting, and forecasting, as well as some financial consolidation and what-if scenario generation. Almost all Centage customers come directly from Excel in the downmarket. Centage competes occasionally against Adaptive

Insights in cases where upper downmarket companies are evaluating multiple first-generation cloud CPM options. The Budget Maestro product is designed as a lightweight Excel replacement to help customers better organize the planning and budgeting process.

The July 2017 release of Budget Maestro 9 included usability improvements to budget models through self-service with its Smart Budget process.

Customers cited Centage's competitive price and improved functionality compared to Excel as primary reasons for choosing Budget Maestro. As far as downmarket vendors are concerned, Centage provides a compelling first step into the CPM world for limited bandwidth customers who want to better organize the way they do CPM processes in Excel.

With the pace of innovation and geographic flexibility of many other cloud vendors, Centage is a niche option for downmarket companies. In 2018, we expect Centage to maintain its current relative usability, while declining somewhat in functionality compared to more mature and broadly relevant vendors. Customers will most likely find value in considering Centage alongside other low barrier-of-entry options when migrating off Excel-only CPM processes.

Host Analytics

Host Analytics was founded in 2001 and is headquartered in Redwood City, California. Host continues to be a leader in the cloud CPM market with high performance in core planning, modeling, reporting, and consolidation functionality. Its data architecture layer, called the PDEX Engine, uses common performance management styles of architecture and API integration. Host leverages Dell Boomi for integration connections to many software systems. The Host platform has robust user security, data governance, and audit management. The Host platform also includes purpose-built modules with specific functionality for various business processes.

In late 2017, Host introduced three new products — Host Dashboards, the Host Model Manager, and the Host Solutions Exchange. Host Dashboards offer visualizations and self-service data analysis. Host Model Manager is a graphical interface within the Modeling module and will likely have positive impact on customer experience with the user interface in 2018. Host Solutions Exchange is a place for customers to discuss best practices. Host has also recently made improvements to the cross-browser and cross-operating system issues facing customers.

Customers reported that the Host user community was very positive and that they felt Host had a strong roadmap moving forward. Most users reported good support, a positive deployment process, and high ease of use. One customer said: "Some systems can be technical—they want all the t's crossed and i's dotted—but the typical average user doesn't have that technical knowledge and just wants to see what they need. Host allows you to configure it this way for them and not get into extraneous stuff." Users also praised the integration capabilities of the product, in particular the use of Dell Boomi. Host partners with notable financial CPM vendors such as BlackLine and Workiva to address areas where it lacks specific accounting functionality.

With its move toward more advanced modeling, additional budgeting and consolidation functionality, and user interface improvements, Host will keep pace in functionality and improve in usability relative to the pace of the market in 2018.

 ## ROI Case Study: Host Analytics

ACI Brands • ROI: 75%

ACI Brands deployed Host Analytics to modernize its financial reporting and budgeting processes. In addition to improving the productivity of the finance department personnel, the company was able to gain better visibility over its customers and product performance, which translated to better decision making and increased profitability. ACI Brands has also implemented Spotlight for Office, which furthers the ease of use for stakeholders in the budgeting process.

THE COMPANY
ACI Brands is a multi-divisional supplier of consumer products serving the Canadian and U.S. retail markets. Founded in 1981, the company specializes in product assortment and category management in an effort to maximize its customers' profitability. Headquartered in the Greater Toronto area in Ontario, Canada, ACI Brands also

has locations in China and the United States.

CUMULATIVE NET BENEFIT

$482,655

$212,436

($264,083)

Year 1   Year 2   Year 3

KEY BENEFIT AREAS

By implementing Host Analytics, ACI Brands modernized its financial budgeting processes. The company did not reduce any costs or reduce headcount, but it was able to increase the efficiency of the finance department personnel and bring better visibility to senior leadership.

Host Analytics allows ACI Brands to eliminate many of the manual and time-intensive tasks that were required of its finance team when it was relying on spreadsheets. Finance personnel were able to redirect to more value-add tasks.

With its prior budgeting process, senior leadership didn't have access to data and reports in a timely manner. Host Analytics allows the finance department to quickly produce the reports they need to generate for company leadership. As a result, executives can get visibility into potential issues and investigate ways to correct them before they become a problem. The company can rationalize poorly performing brands and reduce unprofitable business. ACI Brands can analyze profits and losses and target margins, allowing the business to make pricing decisions or reduce costs.

Finally, with Host Analytics, ACI Brands can more accurately forecast its debt levels. With better debt forecasts, ACI can work with its lenders to achieve more favorable interest rates and reduce the costs it incurs servicing any debt that it carries.

BEST PRACTICES

Trying to do too much too fast was a lesson ACI Brands took away from its implementation of Host Analytics. The initial timeline of the project was overly aggressive. Although ACI was able to get some functionality working very soon after starting the implementation process, the full adoption of the budgeting tools was not possible in such a short timeframe. Ensuring that members of the organization were buying into using the tool was a critical step to help the company realize value from the solution, so devoting enough time for users to adopt the new business processes was necessary.

Since going live with Host, ACI Brands has looked to adopt new features that the vendor is providing to drive additional business value. Specifically, ACI Brands has implemented Host's Spotlight for Office, which extends its reporting capabilities for Microsoft Office. With Spotlight, ACI Brands has continued to make it easier for sales personnel and budget owners to participate in each budgeting cycle. ACI Brands increased the maturity and sophistication of its reporting using Spotlight, bringing additional rigor to its business processes.

· · · · ·

IBM

IBM was founded in 1911 and is headquartered in Armonk, New York. IBM has a large traditional on-premise user base for its legacy Cognos TM1 product. Its cloud products, Planning Analytics and IBM Controller, offer the same user interface in a multi-tenant cloud environment. As an Expert in this Value Matrix, IBM will appeal to specific use cases or customers who want to continue with a "comfort" product that they have experience with.

Users can engage with Planning Analytics through a native Excel interface or through a multi-tenant web interface, which is also available on mobile. The Planning Analytics Workspace allows for user self-service in reporting, dashboard creation, planning, and modeling.

Similar to other large vendors, Planning Analytics will make the most sense for customers with skills and investments in other products from the vendor. In terms of cloud performance management, IBM is a latecomer and customers either were not aware of Planning Analytics or did not see it as a serious candidate for their business during the evaluation stage. It is largely focused upmarket where the mature functionality of its legacy product provides the most value.

Recently, IBM has made improvements around TM1 Server, Planning Analytics Workspace, Planning Analytics for Excel (PAx), and Controller. For the TM1 Server, attribute-based hierarchies, comprehensive object localization, scenario comparison, debugging, server startup, and power Linux conformance have been improved. Workspace has received improvements to planning, analysis, modeling, and operations such as better monitoring and management and enhancements to calculations and formatting. For PAx, shared services, drill-through, autoprovisioning, and publishing capabilities have been improved. IBM Controller has received updates in performance, audit trail, and the Controller Web.

In September 2017, IBM released its most recent product update, including improvements to REST APIs and developer usability. IBM will maintain its position in usability and gain ground in functionality relative to the market as other vendors also advance their cloud products and strengthen their positions as options for legacy customers moving to a first-generation cloud offering.

insightsoftware.com

insightsoftware.com was founded in 2000 and has headquarters in Denver and London. Its CPM product, Hubble, is a highly usable browser-based planning and budgeting tool available on-premise or in the cloud. The product also has functionality for financial close and consolidation and reconciliation. End user roles can be assigned as Explorers (viewers), Power Users, and Designers, as well as for vertical specific roles within sales and finance.

Hubble is designed primarily around integration with JD Edwards and Oracle ERP systems, and it provides a simplified alternative to Hyperion. insightsoftware.com is focusing future innovation on building out greater integration with SAP in addition to its current integration offerings. Hubble customers that we spoke with often previously did their reporting in Excel spreadsheets and rarely considered other vendors on a shortlist before selecting the product.

Recently, insightsoftware.com expanded the breadth of its ERP integration portfolio, specifically focusing on SAP with its integrated replication engine, and will continue in the near future. It is continuing to leverage the 2016 acquisition of Antivia to create better mobile capabilities for dashboarding. In late 2017, Hubble improved integration to SAP integration, improved its mobile dashboarding interface DecisionPoint, and added a new PDF engine.

Hubble customers praised the high usability of the product and excellent customer support. Users also commented that the reporting visualizations were appealing, and that the user interface was overall a good experience. Lastly, the product requires very little IT support and works well for smaller companies that want to limit internal expenses with a lightweight cloud product.

insightsoftware.com has been able to build a happy user base, and we expect it to maintain high user satisfaction as it grows. We see the product keeping pace with the market in terms of functionality and its investments in product improvements.

Jedox

Jedox was founded in 2002 and is headquartered in Freiburg, Germany. Its CPM product incorporates planning, budgeting, and reporting in a single application designed to simplify the user experience and build adoption across core performance management and BI use cases in the office of finance, sales departments, human resources, and other business functions. The product can be deployed on-premise, in the cloud, or as a hybrid solution. The software comes with data governance down to the cell level and its own extract-transfer-load (ETL) tool.

Jedox also partners with Qlik, Power BI, and Tableau to integrate data discovery into its CPM capabilities.

In January 2018, the vendor released Jedox 7.1. The release includes out-of-the-box planning models, improvements to analytics and data visualization, a new mobile interface, improvements to data preparation, changes to the Cloud Console, and a partner solutions exchange called the Jedox Marketplace.

Customers reported that the product was quick and easy to implement, and that competitive price was a major reason for selecting Jedox. The Jedox Excel Add-in offers planning, analytics, and reporting functionality within a native Excel environment. The additional web interface provides customers with a high level of comfort in a browser-based planning environment.

With its largest customer base in the DACH region, Jedox is still expanding in the North American market. Customers also reported that the product has some scalability issues and can be slow when working with large data sets.

Jedox's focus on usability and building out its Excel interface will likely lead to a relative increase in usability in 2018, and we expect its functionality to keep pace with the market, at least until the vendor can establish a greater international presence.

Kaufman Hall

Kaufman Hall was founded in 1985 and is headquartered in Skokie, Illinois. In 2014, it acquired CPM vendor Axiom Software, which continues to provide its performance management product suite. The vendor provides industry-specific functionality for its core focus of financial services, higher education, and healthcare. It provides core CPM functionality around PBF, reporting, and financial close and consolidation. Although the Axiom product has a sizable percentage of on-premise customers, all new deployments are cloud, and Kaufman Hall prioritizes its Azure-based offering. Kaufman Hall is mainly viable for specific industries with need for specialized product features. This is due to its purposefully narrow focus on several verticals, but also because of its geographical spread limited to North America.

We have heard no significant announcements since the last Value Matrix.

Customers found that Kaufman Hall's area expertise helped them in the initial consultation and deployment phases. Customers highly rated the module for Profitability and Cost Management and other industry-specific features. They also reported that online resources were good. One customer said: "Axiom is flexible for import in SQL and ETL, and we've had no performance issues with data sets. That said, we have had some issues when upgrading."

Kaufman Hall is a Core Provider in this Matrix because of its competency in core

CPM capabilities and relevance in specific verticals in North America. We expect Kaufman to maintain its current position in usability and increase in relative functionality as it invests heavily in product capability improvements and continues to push into its chosen verticals.

OneStream Software

OneStream Software was founded in 2011 and is headquartered in Rochester, Michigan. Its product, OneStream XF, offers robust functionality around both accounting and FP&A processes including core PBF, modeling, reporting, financial close and consolidation, reconciliations, disclosure man-agement, and risk management. OneStream also offers a marketplace for specific downloadable solutions. The product is available in the same format as an on-premise or cloud deployment, and it covers a wide range of both corporate and departmental needs and use cases. Although OneStream is a newer vendor, it has demonstrated sustained growth and offers a highly us-able and functional solution. OneStream competes in the mid and enterprise markets for many customers moving to a firstgeneration cloud solution from a legacy on-premise deployment.

We have heard of no recent announcements since the last Value Matrix.

OneStream's breadth of functionality across strategic and financial operations competes well against top vendors in either of these camps. As such, the product offers a strong alternative to the robustness of on-premise Hyperion (or other legacy product) deployments which span a similar range of functions across various modules. In replacing these legacy applications, OneStream has proven its ability to scale to the enterprise level and manage a variety of uses without sacrificing the overall efficacy of its product suite. Customers praised the performance and scalability of the product with large data sets.

On the usability side, customers consistently reported strong vendor support and a positive deployment experience. They noted that OneStream is quick to rollout suggested product improvements and works closely with customers to improve the product around new issues or uses. One customer said: "Everything they have is built on one platform, and to me that is one of the differentiators. What it means is that nothing ever breaks. I've had them implemented for three years and there has been zero downtime, ever—even with the multiple upgrades that have taken place in that time." Customers found the "guided workflow" style interface easy to train on and use and remarked that upgrading is exceptionally easy.

As OneStream continues to build out its functionality to rival that of legacy onpremise vendors', it will improve relative to the pace of the market while maintaining its high level of customer satisfaction and usability.

Oracle

Oracle was founded in 1977 and is headquartered in Redwood City, California. Oracle Hyperion, the vendor's legacy on-premise offering, has long been the dominant product in the CPM market. Several years ago, Oracle made a commitment to a "cloud first" strategy and has since released its EPM Cloud and Planning and Budgeting Cloud Service (PBCS). Oracle continues to operate in the enterprise-scale upper end of the market, bringing years of product maturity and functionality to complex use cases.

The EPM Cloud fits in the market mainly as a replacement to on-premise Hyperion deployments and brings with it similar functionality and reduced complexity, without the need for heavy customization. Overall lower total cost of ownership (TCO) and maintenance requirements make the product appealing to Hyperion customers familiar with the product who are looking to transition to the cloud.

However, Oracle does not address as confidently how its product competes with other cloud vendors who are eating up legacy market share as customers transition to a first-generation cloud solution. Moreover, net new customers tend to consider Oracle because of the name recognition of Hyperion rather than on the proven value of the EPM Cloud. Oracle is somewhat late to entering the cloud arena, and, as a result, is finding itself in competition with many vendors whose go-to-market strategy is to specifically target the Hyperion install base. While the Oracle EPM product suite still provides the most mature and advanced functionality at scale, other vendors are making inroads to offer cloud-based alternatives that rival Oracle's dominance at the upper end of the market.

We have heard of no recent announcements since the last Value Matrix.

Oracle is a standout functional leader with capabilities across all aspects of CPM, including disclosure management, risk and audit assessment, and reconciliations, among more common planning functionality. The product scales to the largest enterprise needs and integrates well with the Oracle tech stack. Customers who used Oracle for ERP and HCM consistently reported good integration and deployments. For its cloud products, Oracle requires little IT resources for end users. Hyperion also has an extensive user community and many online resources to aid users.

Oracle will maintain its position as a functional leader for now, keeping pace with the market of innovations. In the next few years though, Oracle will feel greater pressure from highly scalable cloud CPM vendors.

prevero

prevero was founded in 1994 and is headquartered in Munich, Germany. In July 2016, it was acquired by the Dutch ERP vendor, Unit4. prevero has its strongest presence in the EMEA region but is expanding its footprint in the North American market. Consequently, it specializes in integration and reporting with SAP ERP solutions, but also connects to other major ERPs like Oracle Financials and Microsoft Dynamics. In North America, prevero competes in the midmarket, often against Adaptive Insights.

The prevero product is fully cloud-based on Microsoft Azure, but can be customized for other architectures like the Google Cloud. It provides key functionality around PBF, predictive analytics, and risk management, and is building out greater financial functionality in close and consolidation. In the near future, prevero is targeting the verticals of professional services, higher education, and non-profit. Like some other CPM vendors, prevero incorporates both planning and BI within a single product, aiming to get a more diverse user base within its customers and to enable those users to manage multiple business tasks with one tool.

We have heard no recent announcements since the last Value Matrix.

prevero has a solid mix of strategic planning, financial operations, and business intelligence functionality within a single application that could prove valuable to companies with relatively simple requirements in any of these three camps. Exposure to a range of functionality—in conjunction with the Power User model of deployment favored by prevero users—will likely lead to greater user adoption of multiple parts of the tool in companies where such functionality is consolidated into a small number of employee roles (for example, where the FP&A team also runs BI reporting). One customer said: "The drag-and-drop graphical user interface is super easy for end users, although for central controller (power user), it's is much more complicated. Support is very good, community resources are good, and the implementation was easy."

prevero works well for companies that fall into its vertical expertise or with interest in expanding the FP&A role into more traditional BI within a single application. We expect prevero to grow its North American footprint over the next year and offer more value to customers as an alternative to established cloud CPM products. As such, prevero will keep pace with the market in terms of usability and functionality in 2018.

Prophix

Prophix was founded in 1987 and is headquartered in Mississauga, Canada. The vendor offers the choice of on-premise and cloud solutions with functionality primarily in PBF, FCC, and financial reporting. Both deployment options offer the same user experience. Prophix also offers analytics—including self-service dashboards and visualizations—as well as functionality for data integration, workflow, and process automation. Prophix's customers span all industries and are primarily located in North America, the UK, Europe, and South America. Many of Prophix's customers come straight from Excel, and most frequently considered Oracle, Adaptive Insights, and Host Analytics before selecting Prophix. Many of its customers chose the solution for its lower total cost of ownership.

In May 2017, Prophix announced a strategic partnership with ERP provider SYSPRO and in January 2018, Prophix furthered its partnership with ERP vendor Sage. SYSPRO and Sage customers now have access to Prophix's CPM solutions and can integrate it with their existing ERP data. In the past year, Prophix has launched several updates to its visual user experience for model management, template building, workflow management, and other administrative capabilities.

Customers note that Prophix is easy to deploy and requires little ongoing maintenance. Prophix has relatively high usability for a core provider because it has a visual web-based interface. End-users also report that it is easy to integrate data and that Prophix provides strong customer support.

Prophix does not offer functionality for compliance and risk management. Additionally, customers note that the solution can sometimes be slow in running reports and that it is hard to toggle between information in the solution.

In 2018, we expect that Prophix will continue to gain new customers across the Americas and Europe. In the next year, we see them keeping pace with the market in both usability and functionality.

ROI Case Study: Prophix

Lochmueller Group • ROI: 95%

Lochmueller Group deployed Prophix to improve the efficiency and accuracy of its financial budgeting. By leveraging the flexibility of Prophix's corporate performance management (CPM) solution, Lochmueller Group fits the software to its business needs. In addition to reduced time spent on year-end budgeting, the company has experienced increased productivity from its branch managers and discipline leads with Prophix making the budgeting process more collaborative.

CUMULATIVE NET BENEFIT

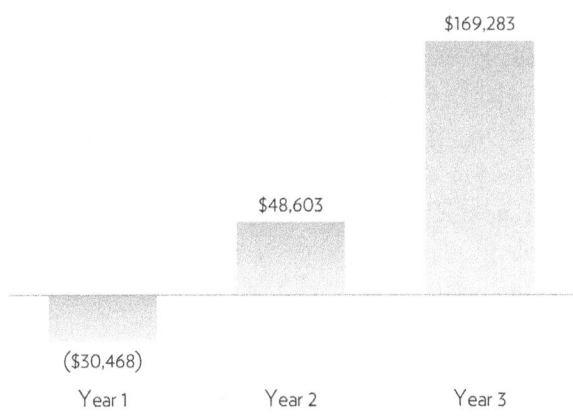

		$169,283
	$48,603	
($30,468)		
Year 1	Year 2	Year 3

THE COMPANY

Lochmueller Group is an engineering firm that specializes in providing transportation, planning, and environmental services, in addition to data collection and modeling. Headquartered in Evansville, Indiana, Lochmueller Group has been in business since 1980. The company has about 180 employees and built a name for itself by balancing its planning and engineering efforts with a concern for the environment and the ability to provide clients with practical and cost-effective designs.

KEY BENEFIT AREAS
Deploying Prophix improved Lochmueller Group's financial budgeting processes and reduced the amount of time necessary to complete them.

- Prophix reduced time spent on year-end budgeting and analysis. Lochmueller Group branch managers and discipline leaders can complete their year-end processes and data analysis more quickly with Prophix.
- Prophix also increased management productivity. With better access to the data they need, management is more productive and can spend less time on budgeting, quarterly forecasts, and monthly reporting.
- Finally, Prophix helped reduce report creation and distribution time. Employees are able to more efficiently create and distribute reports at month- and year-end.

BEST PRACTICES
Lochmueller learned from its experience configuring Prophix to handle the complexity of its interoffice transactions that the solution's flexibility is one of its primary differentiators. The company knows it can rely on Prophix being able to adjust to changes in its budgeting and forecasting needs. Lochmueller has already identified ways in which it will build out Prophix's capabilities and derive more value from the solution moving forward.

The company leveraged the functionality of Prophix to make the budgeting process more collaborative between the budget owners and the accounting department. With better access to information, managers can improve their decision-making when creating their budgets and adding resources.

· · · · ·

SAP

SAP was founded in 1972 and is headquartered in Walldorf, Germany.

SAP has a large legacy install base of its on-premise offering, SAP Business Planning and Consolidation (BPC). SAP's cloud offering is the SAP Analytics Cloud, which is available on HANA One or AWS (Amazon Web Services). SAP has traditionally competed with other legacy on-premise vendors in the enterprise market and has struggled to shift to a competitive cloud alternative. A latecomer to the cloud game, SAP — like Oracle and IBM — has had most of its cloud success with its existing install base, articulating a value message around migration and providing on-premise customers with a roadmap to the cloud that often leverages the flexible hybrid deployment options available with the Analytics Cloud.

That said, SAP remains an Expert in the Value Matrix because of its mature functionality across all CPM capabilities. BPC and the Analytics Cloud are realistic options for current SAP ERP customers, but SAP does not often show up on short lists for net new customers.

In early 2018, SAP published the latest release of SAP Analytics Cloud, which included functionality updates to story dashboards and planning and usability updates for the modeling interface. Late 2017 releases also included many improvements to stories, planning, modeling, and integration.

Customers remarked that the product was very robust with excellent modeling and reporting drill-down and slice and dice features, for example. Overall, the SAP products have strong functionality that scales well for enterprise-sized data requirements, and they are flexible enough to accommodate many different strategic planning and budgeting processes. SAP also has good functionality around accounting processes in conjunction with an SAP ERP system. Customers liked the Excel-style front end for the FP&A team.

SAP will largely retain its legacy install base over the next several years, although the number and speed of customers that will transition to the cloud is unclear. Net new customers looking for a first-generation cloud offering might place SAP on their shortlist due to name recognition, but we found that customers who chose SAP usually were already invested in the SAP product stack. SAP will decrease in relative functionality to the market as other products catch up to the maturity of the SAP product suite.

Vena Solutions

Vena Solutions was founded in 2011 and is headquartered in Toronto, Ontario. The Vena product focuses mainly on core planning, budgeting, forecasting, reporting, and financial close management for midmarket and enterprise customers. Vena delivers a multi-tenant cloud solution that is web and mobile compatible and fully integrated with Microsoft Office 365 Online and desktop.

Vena also brings a vertical-specific expertise to financial services and manufacturing customers, and provides specialized implementation, training, and consulting support for customers in these industries. Vena recently launched the Vena Exchange, a platform where users can share input and reporting templates and best practices.

The Vena product requires little IT support for individual business units like the FP&A or accounting teams. Although Vena is primarily used for its planning and budgeting functionality, some customers use it strictly for close and account reconciliation, or in combination with PBF. As a result, Vena competes with a range of cloud vendors in a number of use cases.

In 2017, Vena focused on expanding its range of data integrations from external sources such as NetSuite, Salesforce, QuickBooks Online, Zapier, and Intacct. Vena has also recently introduced Revenue Performance Management for improved resource allocation for subscription-based businesses.

Vena's diverse functionality allows customers to employ the product for several different departments within the organization. Usability was the high point noted consistently by customers: the product is easy to use and deploy with a fast learning curve and has low IT needs. Customers reported that vendor interactions were top notch. Support is extremely responsive and incorporates suggested improvements rapidly. The sales process was also smooth, and price was a strong selling point for many customers. One customer said: "We went through an extensive vendor selection process — we went through every possible vendor who could meet our performance requirement and multiple vendors did demos. Then we selected three vendors to do a proof of concept and selected Vena because the users really liked being able to work in Excel. Once the templates were there, it was easy to use."

We see Vena improving in functionality in 2018 as it builds out its nascent capabilities across core strategic and accounting aspects of the product. In tandem with these improvements will likely be an increase in customer wins where Vena checks multiple boxes on customer functionality requirements. Vena is a newer vendor in the marketplace and will need to build out its user base — both through new customer wins and through departmental expansions within its current customers — before it can articulate a multi-department value message that will make it a company-wide solution at the proof-of-concept stage.

Workiva

Workiva was founded in 2008 and is headquartered in Ames, Iowa. The company's roots are in SEC filing software with a user interface similar to that of Microsoft Office products. However, Workiva's cloud-only product, Wdesk, also provides functionality for financial reporting, performance reporting, Sarbanes-Oxley (SOX) compliance; data management and aggregation, governance, workflow management, dashboards and reports, and some PBF.

As a facilitator in the CPM market, Workiva dominates the market in SEC reporting and disclosure software. In this space, its only competitors are traditional financial printers such as RR Donnelley and Merrill Corporation. A majority of Workiva's customers use the product for SEC reporting. Many of those customers also use Wdesk for financial and performance reporting and broader CPM capabilities. On the more traditional PBF side, Workiva mainly competes with Microsoft Excel and addresses specialty PBF use cases in both large and smaller customers.

In May 2017, Workiva partnered with Host Analytics. Workiva integrated Wdesk with the Host Analytics EPM platform, giving Workiva customers functionality in finance and accounting including extended PBF as well as consolidation and financial reporting. In July 2017, Workiva announced data integrations between Wdesk and over 100 cloud, SaaS, and on-premise applications including Oracle ERP.

Workiva is best known for its SEC filing software. One of Workiva's main strengths is its cloud data management capabilities, namely its ability to import and aggregate data from Excel documents, ERP, CRM, and other documents and systems. Workiva's native eXtensible Business Reporting Language (XBRL) functionality enables end users to tag all financial elements in the preparation of an SEC filing, track the cell-level history of every number, and ensure accuracy in financial reporting. Workiva's workflow management and collaboration capabilities are also strong. Users can set workflows for structured and repeatable processes, and from this workflow they can assign tasks and permissions to specific users who can collaborate in real time. End-users also report that Workiva has strong customer support. One customer said: "Everything we've done has been quite easy. It's cloud-based so everything just turns on. It's user friendly. For the most part, we do the implementation of any new processes ourselves."

In 2018, we expect to see Workiva expand its data preparation capabilities to provide users with expanded functionality to import, consolidate, cleanse, and map data, facilitating end users' ability to prepare financial reports. We see the company keeping pace with the market in terms of functionality and usability in 2018.

Marketing Automation Value Matrix

Dot indicates current position. Arrow indicates trend in 2019 relative to others in the market.

CHAPTER 12
MARKETING AUTOMATION

Beyond the marketing hype of the connected customer experience, marketers are now challenged more than ever to reach the holy grail of digital marketing: zero opt outs. Marketers approach this goal when they balance hyperpersonalization with sensitivity, when they see marketing as only one channel of many to reach customers, and when they can pull data and insights in real time from within the other core pillars of CRM (service and sales), as well as commerce interactions, to inform their outreach. As a result, leaders in the Matrix have invested heavily in integration, AI and analytics, and data capabilities to help their users in the quest for the holy grail. Other trends impacting the marketing automation space include:

General Data Protection Regulation (GDPR). Concerns about data privacy are top of mind for both consumers and business buyers and are only exacerbated by the European General Data Protection Regulation (GDPR) requirements and deadlines, which, in theory, make it easier for consumers to opt out of digital marketing. No field has been hit harder than the European Union's GDPR requirements than marketing, and vendors are providing both new capabilities to support compliance and guidance for customers on how to ensure it. The GDPR conversation has shifted from "who needs to be compliant and when" to "how do I apply these principles to the way I manage all my customer data to increase transparency and trust?". Companies that are not already GDPR compliant are under the gun to not just reach compliance but to deliver the same transparency and trust around consumer data as their peers.

- **Personalization.** Every marketing automation vendor talks about personalization, but few customers have achieved success: beyond even microsegmentation, true personalization means no prospect or customer ever receives communication they don't want, at a time or from a channel they don't prefer. Once a person opts-out, it's even more difficult now with GDPR to re-engage them with marketing, so effective personalization is critical.
- **Blurring lines between business-to-business and business-to-consumer marketing.** With more and more business buyers looking to the Internet to research, price, and even buy products without interacting with a sales person, marketing tools that were once clearly delineated as business-to-business or business-to-consumer tools are taking on blended characteristics.
- **Consolidation and integration.** One of the biggest hurdles to effective omnichannel marketing is real-time visibility into data across multiple departments, databases, and applications — and vendors are driving both industry consolidation and deeper integration within their acquired assets to deliver on the "single view" for both existing customers and prospects.

A key component of this strategy is, not surprisingly, tight integration with sales force automation (SFA), enabling users to measure end-to-end marketing automation impact on actual sales and provide micromarketing and lead scoring capabilities for greater insight and execution for sales teams.

Nucleus sees commerce integration as important as well, with many vendors promoting e-commerce and marketing integration (triggered marketing to an individual that abandons a shopping cart is a simplistic example that is becoming table stakes).

· · · · ·

ActiveCampaign

ActiveCampaign provides e-mail marketing, marketing automation, sales automation, and other core CRM functionality for SMBs with a focus on bloggers, agencies, software as a service companies, and higher education. Its marketing automation capabilities include campaign management, lead management, social, mobile, and reporting. ActiveCampaign has made investments to support clients' GDPR compliance including improving site tracking features, integrating with GDPR compliance capabilities from BigCommerce and Shopify, and improving contact deletion capabilities to comply with "right to be forgotten requests."

Act-On

Act-On caters to mid-market enterprises, providing a marketing automation platform with capabilities including e-mail marketing, social media marketing, and content marketing. The company's differentiation is its focus on the concept of adaptive journeys, with adaptive segmentation and forms, and adaptive scoring, sending, and channels to support data-driven engagement optimization. The company has seen significant changes in the past six months, with a number of new C-level appointments including CEO, CTO, and CRO, and more than 80 new hires in the first six months of 2018.

Adobe

Adobe Marketing Cloud is an integrated digital marketing platform comprised of 9 marketing tools: Analytics, Audience Manager, Campaign, Experience Manager, Media Optimizer, Primetime, Social, Target, and TubeMogul. This year, Adobe announced new intelligent content and integration capabilities for Adobe Experience Manager including image discovery and tailoring, automatic personalization of content and form layout, and integration of assets between Experience Manager and Creative Cloud applications. Additionally, it has invested heavily in machine learning and artificial intelligence with new capabilities to predict the best images for e-mails and predict customer churn, as well as new Adobe Sensei-based capabilities for Adobe Target to support more personalized experiences. In 2018, Adobe acquired Marketo to bolster its marketing automation capabilities.

Agillic

Agillic is a Denmark-based marketing automation provider that differentiates itself with a centralized customer database and personalization engine that leverages AI and predictive analytics across customer journeys with support for channel and time optimization, targeting, and personalization. The solution supports personalization and data collection through e-mail, print,

Web, app push, text and SMS, landing page, social, forms, and beacons. The Agillic Portal also supports contact center agents with a full view of customer interaction history.

Bridgeline Unbound

The Bridgeline Unbound Digital Experience Platform is a unified suite of content, marketing, commerce, social, and insights components for managing digital experiences across multiple channels. Bridgeline also offers a core accelerator framework of flexible templates and modules for rapidly implementing digital experiences which provides customers with cost effective solutions and the benefit of velocity to market.

Founded as an enterprise content management provider, the company has evolved its capabilities to an impressive, broad suite of products — integrating marketing automation, rich e-commerce functionality, social media management as well as search engine optimization. The Bridgeline Unbound solution allows organizations to efficiently manage large hierarchies of localized sites, campaigns and automation flows. The solution is offered as a cloud-based, multitenant environment in addition to managed services.

Engagio

Engagio provides cloud-based account-based marketing (ABM), campaign management, and marketing orchestration solutions, taking an ABM-centric approach to supporting business-to-business marketers. Some key capabilities of the product include "Dash Account Based Attribution," a new product designed to help marketers understand the impact of marketing programs on multiple people within an account and evaluate results using multi-touch attribution models, and expanded capabilities for measuring and defining marketing qualified accounts (MQAs), with filters for minutes, people, actions, and time frame, and an exclusion list to enable marketers to automatically exclude certain accounts from being considered as MQAs.

GetResponse

GetResponse is positioned as an all-in-one online marketing platform focused for SMBs, with capabilities for e-mail marketing, Webinar marketing, landing pages, marketing automation, and CRM. Its e-mail marketing features include customization and personalization, advanced segmentation, optimization, tracking, and testing. Targeting SMB customers, the product is easy-to-use with customizable templates for the most common marketing scenarios, and the GetResponse Marketplace serves as a platform to connect business experts in marketing, sales, and design with GetResponse customers.

HubSpot

HubSpot Marketing has its roots in inbound content marketing focused on SMBs, and its solution includes marketing automation; e-mail marketing; campaign, content, and lead management; Web development and hosting; search engine optimization; and social capabilities. The company also provides free service automation and sales automation modules. The product connects with popular social media sites to enable social marketing through new tools inside Marketing Hub including a new Facebook lead ads integration, Facebook ads audience sync, and Instagram integration. This year, the company also announced Conversations, a tool to help teams manage one-to-one communications across their Web site pages, Facebook, Slack, and other messaging channels.

IBM

IBM has rebranded its Marketing Cloud, Campaign and Analytics assets under the umbrella of IBM Watson Marketing. IBM markets this portfolio around three key pillars: personalized marketing, customer insights, and digital experience. Personalized Marketing includes Watson Campaign Automation, Real-time Personalization and the IBM Marketing Software portfolio (which includes Campaign, Interact, Marketing Operations, and Contact Optimization). Customer Insights includes Customer Experience Analytics, Marketing Insights and Customer Insights. Digital Experience includes the Digital Experience portfolio and Watson Content Hub. In early 2018, Watson Marketing rounded out the portfolio by adding Media Optimizer, a solution that includes a DSP and DMP. Underpinning the Watson Marketing solutions is Universal Behavior Exchange which integrates both IBM and third-party solutions across the marketing ecosystem.

Infor

Infor Omni-channel Campaign Management includes capabilities for campaign management, e-mail marketing, customer targeting and segmentation, database management, Online Analytical Processing (OLAP), and predictive analytics. Historically, Infor has had deep functionality and capabilities for personalization at scale; however, recently it has been focused more on delivering marketing capabilities through its partnership with Marketo.

Infusionsoft

Infusionsoft provides an integrated marketing automation, customer relationship management, and e-commerce solution for small businesses and entrepreneurs. In May, the company announced a reengineered product to be rolled out over the next several months which includes industry-specific templates for e-mails and landing pages, a new visual sales pipeline, a redesigned mobile experience, Infusionsoft Payments, and access to on-demand marketing coaches and 1,500 Infusionsoft partners with industry-specific expertise.

Mautic

Mautic provides marketing automation both through a free open-source product and through a hosted product known as Mautic Cloud. Capabilities of the Mautic solution include contact management; customer lifecycle management; multi-channel marketing; reporting; and lead scoring, nurturing, and conversion. In October, the company announced Maestro, a marketing management solution for agencies and enterprises that enables users who need to manage multiple client accounts, supporting campaign replication, custom branding, and benchmark reporting. The company also announced a beta program for Maven Marketing Intelligence, an AI-driven campaign intelligence tool that supports automation based on analysis of historical data on best practices.

Microsoft

Microsoft Dynamics 365 for Marketing makes its first appearance in the Marketing Value Matrix. The product was released in Spring 2018 and provides marketing automation capabilities that are tightly integrated with Dynamics 365 for Sales and has built-in business intelligence capabilities. The initial release provides capabilities for users to create graphical e-mail messages and online content to support marketing initiatives, design interactive customer journeys, leverage LinkedIn data for leads, organize and publicize events, and analyze marketing results and ROI.

Oracle

Oracle Marketing Cloud includes Oracle Eloqua, Oracle Responsys, Oracle Maxmiser, Content Marketing, Social Marketing, and Industry Solutions. Oracle differentiates its offerings with a data focus, specifically with its integration to its Data Cloud offering—which is not surprising given Oracle's roots as a database company. In a recent analysis of Oracle customers, we found that retailers moving to Oracle Marketing Cloud were able to increase marketer productivity by an average of 50 percent while increasing customer engagement, leading to an increase in campaign-driven revenues of 20 to 30 percent.

This year, Oracle also announced a suite of new capabilities powered by AI called Adaptive Intelligent Apps for CX. These are embedded within the Oracle CX Cloud Suite to provide what Oracle calls "Connected Intelligence" not only across CX but across the business, including in ERP and HCM. Connected Intelligence shares and coordinates data and data insights across platforms and channels.

Salesforce

Salesforce's marketing automation capabilities are comprised of two main offerings: Salesforce Journey Builder, Salesforce's consumer engagement solution, which is part of the corporate and enterprise e-mail, mobile, and Web marketing editions of Marketing Cloud, and Salesforce Pardot, Salesforce's business-to-business marketing automation solution, which includes e-mail marketing, lead generation and management, ROI reporting, and sales alignment.

These two core products are supported by the Salesforce platform and Salesforce Einstein, which delivers AI-driven recommendations and insights across Marketing Cloud. Recent investments have been focused on expanding the reach of the product, shown by a new integration between Google

Analytics 360 and Salesforce which will allow marketers to create audiences in Analytics 360 and then activate those audiences for engagement within Marketing Cloud. Another key investment area is intelligence, demonstrated by updates such as Einstein Segmentation (part of the Salesforce data management platform (DMP), which analyses consumer signals to discover new audiences to reach with personal campaigns, and Einstein Splits, enabling marketers to create unique personalized journey paths for each customer.

 ## ROI Case Study: Salesforce

VMWare Inc. • ROI: 641%

VMWare Inc. (VMW) deployed Salesforce Pardot, Marketing Cloud, and Sales Cloud to facilitate a new business-to-business (B2B) marketing strategy for its cloud portfolio, and to integrate its marketing data with sales to produce a more unified customer view. The project enabled VMW to increase user productivity, accelerate campaign launch times from months to weeks, and increase profits through improved lead generation and conversion.

CUMULATIVE NET BENEFIT

		$30,787.385
	$19,161.885	
$8,736.385		
Year 1	Year 2	Year 3

THE COMPANY
VMW is a publicly-traded subsidiary of Dell Technologies, based out of Palo Alto, California, with more than 120 locations and 20,000 employees worldwide. Founded

in 1998, VMW was acquired by EMC Corporation in 2004 (which Dell acquired in 2016) for its platform virtualization capabilities. It continues to develop and sell cloud computing and virtualization products with the aim of separating application software from its underlying hardware to create a flexible digital foundation to support businesses.

KEY BENEFIT AREAS

Deploying Salesforce Pardot, Marketing Cloud, and Sales Cloud together allowed VMW to modernize its sales and marketing strategy, decrease spending on IT support, and increase total sales. Overall agility was improved, and the company can now execute at speeds that were impossible on the legacy system, driving increases in productivity and profits.

- **Eliminated legacy system support costs.** Moving to Salesforce allowed VMW to retire legacy software and achieve savings from eliminated license fees. The simplicity of the new stack reduced the need for ongoing personnel support; subsequently one full-time IT staffer was redeployed.
- **Increased contributor productivity.** On-platform automation, the usability of the platform, and sales and marketing data unified on the Sales Cloud have increased contributor productivity by 20 percent. Automated data generation removes human error and improves data quality with Einstein Automated Contacts. On the legacy system, a single ad campaign took six months to build, while on Salesforce this is reduced to two and a half weeks.
- **Increased marketing effectiveness.** With the ability to turn out campaigns at a faster rate, VMW better able to segment and target the right customers. With analytics capabilities from Salesforce Einstein embedded, teams can leverage sales, marketing, and customer data to sell more effectively. For lead generation, the project allowed

VMW to market a cohesive message across all 17 products and services; in the first day of go-live, sales met and doubled its lead target for the month.

- **Additional revenue capture through marketing.** Centralizing all customer data allows users to perform lead scoring and lead prioritization and deliver the insights to sales teams. Marketers have access to recent sales data, allowing them to create customized campaigns to more effectively leverage account-based marketing. Twenty percent of additional revenues are attributed to the deployment.

BEST PRACTICES

This case is a clear example of a multinational corporation identifying a sunk cost in an inefficient legacy system and embracing innovation on the cloud at the enterprise scale. VMW realized the need to integrate sales with marketing to facilitate the comprehensive customer understanding needed that drives sales in the modern marketplace. To reconfigure the legacy system would have been costly in personnel time and lost productivity, so investment in cloud innovation was the more feasible choice, particularly given the benefits to collaboration, data management, business agility, and total ROI that we have identified as characteristic to cloud deployments.

Within VMW, the deployment was smoothly executed, taking only three and a half months from start to finish. A crucial policy for companies undertaking a platform migration on this scale is to provide adequate training on the system in order to ensure end-user confidence and adoption.

· · · · ·

SAP

SAP's offerings in this space include SAP Marketing Cloud, a suite of cloud marketing solutions that includes Commerce Marketing; Dynamics

Customer Profiling; Loyalty Marketing; Marketing Planning and Performance; Lead Management; Marketing Analytics; and Segments, Campaigns, and Journeys, as well as LeadRocket, CallidusCloud's marketing automation solution focused on marketing and sales alignment, automating marketing processes, and accelerating sales. Key areas of functionality include e-mail marketing; segmentation; analytics; lead generation, scoring, nurturing, and alerts; and sales acceleration.

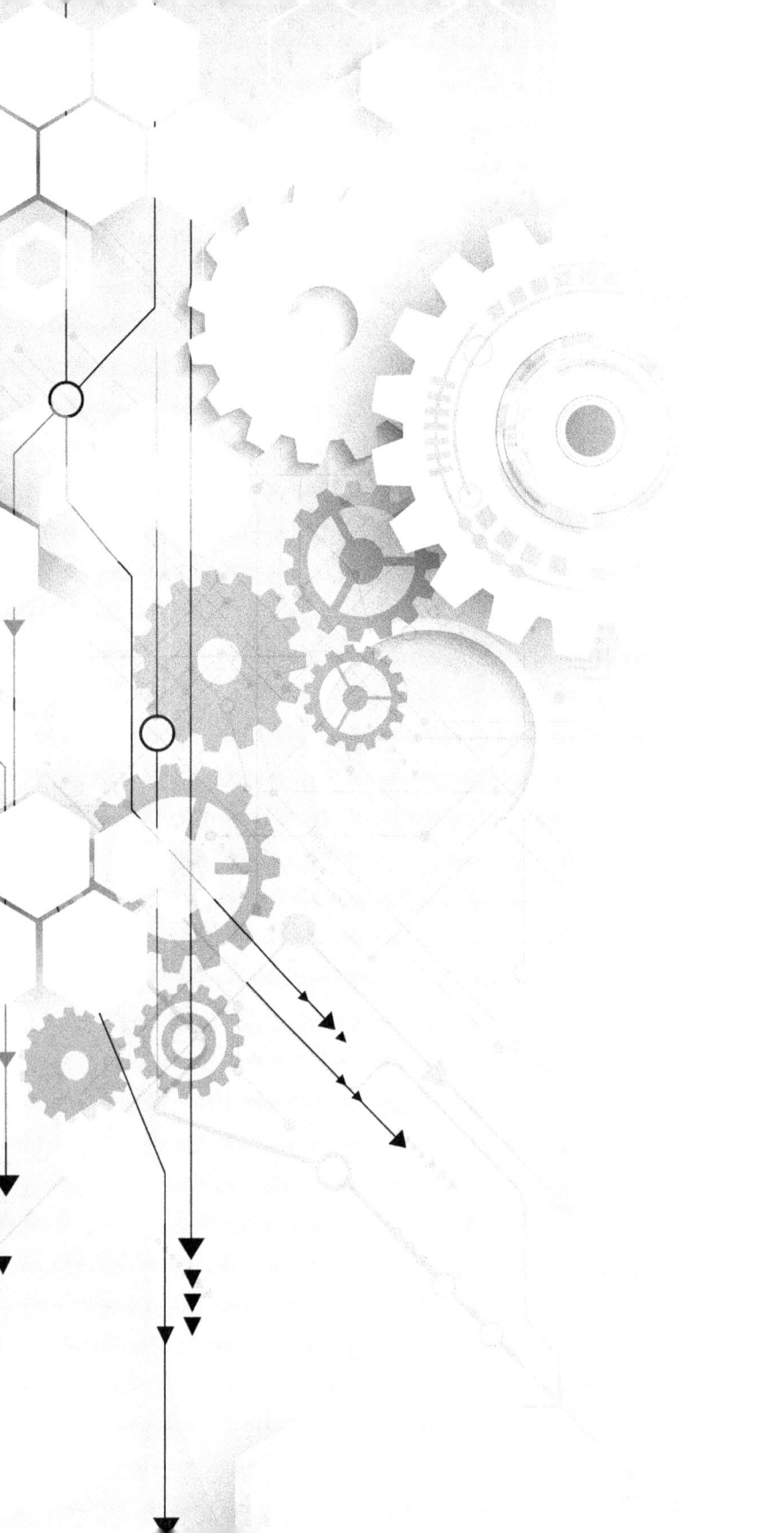

CHAPTER 13
CONCLUSION

I hope this book has proven to be a valuable resource in making sound tech decisions based on numbers and real data as opposed to further peddling of the usual media hype and social buzz. Beyond the information and tools for each enterprise solution area in this book, there are some bigger trends that will drive the greater tech market going forward.

Integration is the mega-trend that we are seeing grow year-over-year. Over the past few years, we've seen great value from the integration of core applications, such as CRM and ERP. That trend continues with even deeper integration and new connections with other application areas, such as the supply chain. A few years ago, I wrote a blog about how everything was moving toward one big "mega app" and that trend has only continued.

Take HCM, an application that has primarily been an HR tool used primarily by staff within that department. Today we are seeing HCM become more integrated with core operational and even finance applications such as ERP and CPM. And while HCM now connects with business units beyond HR, it is also consolidating other tools for a more complete HR package. Talent Management—one of the hottest apps in the current tight labor market—has become a prime M&A target for HCM vendors. And WFM is coming up on the horizon. Tomorrow's HCM is vastly more valuable with greater functionality and usability, leveraging data across other units.

In fact, HR applications are following a similar path to analytics, which has gone from a stand-alone application to primarily an embedded solution.

This is all part of the integration trend toward the "mega app."

Broader functionality with niche capabilities is another big trend emerging. We've seen where integration enables better usability, but it's also driving new functionality as vendors leverage broader access to data and other tools. CPM vendors, for example, are able to go beyond financial reporting and analysis to offer tax provisioning, lease accounting and transfer pricing solutions now.

That includes vertical specialization as some vendors add features that help organizations in the pharmaceutical, manufacturing or financial services industries. Some are even able to target narrow micro-verticals, taking broader functionality to a specific niche industry. That's only making the overall enterprise software landscape more competitive.

Machine Learning and AI are another big trend to watch in 2019. I know I've called out AI as being overhyped. That's because most of the AI coverage is still about distant capabilities further out in the future. Star Trek might have accurately predicted many technologies that have eventually come to fruition, but they were pie-in-the-sky predictions at the time that took decades to pan out as part of the long game. Getting excited about what AI can do in 10 years' time is fun and certainly something to consider for long-term planning. But for 2019, you need to focus on the tangibles.

Machine learning and early applications of other forms of AI for business are indeed starting to make an impact. The biggest benefit we see for 2019 is where AI starts to automate menial tasks that allow employees to focus on more strategic work. The ability to automate forms and data collection, for example. Machine learning is driving new trends, such as micro-coaching for CRM to enable better customer experiences. The potential for training is massive.

It's worth asking vendors about machine learning and AI capabilities within current offerings and what they expect over the next 18 months. Anything beyond that is good-to-know, but not all that relevant to current business decisions. (Especially since so many solutions are now delivered via a Software-as-a-Service SaaS model, which are easy to switch as another vendor offers more value.)

There are other trends that will emerge as the year progresses, but these three are the big trends that we see growing over the foreseeable future:

Integration, broader functionality with niche capabilities and machine learning or AI. You'll notice we haven't made much mention of cloud computing. That's because the cloud revolution is all but complete. Sure, there are still some on-premise applications, but they are in the minority and will continue to wane as a percentage. ERP, with lengthy upgrade cycles, was late to the game but has primarily shifted to the cloud now, as has much of the supply chain. CPM seems to be the last big hold-out, which is not surprising given the reluctance many businesses have had in putting financial data into the cloud.

That being said, simply having a cloud solution is no longer an advantage. It puts the vendor on the field and not much more. Now it's all about how the vendor leverages the cloud to improve its offering to customers. Look closer at how each vendor delivers the functionality and usability your business requires, at the best price.

Make Better Decisions

Decide if you get excited by future capabilities that feel more like a sci-fi movie, or the idea of adding real value to your business. The former is highly entertaining, no doubt. But the later could save your job and even get you a promotion.

So, it's your choice. Invest in shiny objects that will likely be last year's news by the end of 2019, or technologies that can help attract and retain top talent. Forge deeper, more lucrative connections with your customers. Accelerate your supply chain and drive up profits.

Nucleus has been evaluating tech value for nearly two decades. We've helped numerous companies add significant value and improve their business through smarter tech decisions. Keep this book to help you assess technology as needed throughout 2019, and stay tuned for new research throughout the year.

GLOSSARY OF ABBREVIATIONS

ACA	Affordable Care Act
AEC	architecture, engineering, and construction
AI	artificial intelligence
ATS	applicant tracking system
AWS	Amazon Web Services
BI	business intelligence
BPM	business process management
CM	content management
CPA	certified public accountant
CPM	corporate performance management
CPQ	configure, price, quote
CX	customer experience
DaaS	data as a service
EAM	enterprise asset management
EMEA	Europe, the Middle East, and Africa
ERP	enterprise resource planning
ETL	extract, transform, and load
FLSA	Fair Labor Standards Act
GDPR	General Data Protection Regulation
GL	general ledger
HCM	human capital management
IBP	integrated business planning
IO	inventory optimization

IoT	internet of things
ISV	independent software vendor
KPI	key performance indicators
LMS	learning management system
MEIO	multiple echelon inventory optimization
MES	manufacturing execution system
MIO	multi-enterprise inventory optimization
MSS	manager self-service
NLP	natural language processing
OBIEE	Oracle Business Intelligence Enterprise Edition
OCR	optical character recognition
OEM	original equipment manufacturers
OLAP	online analytical processing
PaaS	platform as a service
PBF	planning, budgeting, and forecasting
PEO	professional employer organization
PSA	professional services automation
RFP	request for proposal
ROI	return on investment
RTVN	real time value network
S&OP	sales & operations planning
SaaS	software as a service
SCM	supply chain management
SFA	salesforce automation
SKU	stock keeping unit
SMB	small and medium business
SOA	service-oriented architecture
SPM	sales performance management
SQL	sales qualified leads
TCO	total cost of ownership
UI	user interface
UX	user experience
VAR	value added reseller
WFM	workforce management

INDEX

ABOUT THE AUTHOR

Ian Campbell is the Chief Executive Officer of Nucleus Research where he is responsible for the company's investigative research approach, product set, and overall corporate direction. He is a recognized expert on the return on investment (ROI) and total cost of ownership (TCO) analysis of technology and has written and presented extensively on a range of organizational topics and the importance of matching technology to business organizational objectives. As an expert on technology value, he is a frequent speaker at industry and business events and has been quoted in major business publications including *The New York Times*, *The Wall Street Journal*, *The Economist*, and was the subject of a personal profile in *The Financial Times*.

For over a decade, Mr. Campbell has taught a course on assessing the value of technology at Babson College in Massachusetts and is a frequent guest lecturer at Stanford University, the University of California at Berkley, Massachusetts Institute of Technology, Harvard University, and Boston College.

In addition to his expertise in financial analysis, he is noted for his research identifying the human barriers to a successful technology deployment and the strategies that can be employed to maximize user acceptance of new technology.

Prior to joining Nucleus Research, Mr. Campbell was the Vice President at International Data Corporation where he managed a portfolio of research programs in the US and Europe.

Mr. Campbell holds a Bachelor of Science degree in computer science and economics from Northeastern University and a Masters degree in business administration from Babson College.

ABOUT NUCLEUS RESEARCH

Nucleus Research is a global provider of investigative, case-based technology research and advisory services. We deliver the numbers that drive business decisions.

For more information, visit NucleusResearch.com or follow us on Twitter at @NucleusResearch.

100 State Street · Boston, MA 02109 · +1 617-720-2000
www.NucleusResearch.com

www.ingramcontent.com/pod-product-compliance
Lightning Source LLC
Chambersburg PA
CBHW071254220526
45468CB00001B/127

* 9 7 8 1 7 9 3 9 3 0 0 4 0 *